T0282636

# CAMBRIDGE LIBRARY COLLECTION

*Books of enduring scholarly value*

## Botany and Horticulture

Until the nineteenth century, the investigation of natural phenomena,
plants and animals was considered either the preserve of elite scholars or a
pastime for the leisured upper classes. As increasing academic rigour and
systematisation was brought to the study of 'natural history', its subdisciplines
were adopted into university curricula, and learned societies (such as the
Royal Horticultural Society, founded in 1804) were established to support
research in these areas. A related development was strong enthusiasm
for exotic garden plants, which resulted in plant collecting expeditions to
every corner of the globe, sometimes with tragic consequences. This series
includes accounts of some of those expeditions, detailed reference works
on the flora of different regions, and practical advice for amateur and
professional gardeners.

## The Uses of Plants

An eminent botanist and natural historian, George Simonds Boulger (1853–
1922) wrote a number of books on plant life in the British Isles. He published
this concise work in 1889. It opens with a brief survey of the progress made
in economic botany over the years, particularly in the period 1837–87. Boulger
then notes the commercial application of plants across many fields, notably
food production, medicine, and the building trade. Common and botanical
names are given, followed by succinct descriptions of each plant. Including
both a general and synoptical index, this accessible resource can be read with
profit alongside John Jackson's *Commercial Botany of the Nineteenth Century*
(1890) and Boulger's *Wood: A Manual of the Natural History and Industrial
Applications of the Timbers of Commerce* (1902), both of which are reissued in
this series.

Cambridge University Press has long been a pioneer in the reissuing of out-of-print titles from its own backlist, producing digital reprints of books that are still sought after by scholars and students but could not be reprinted economically using traditional technology. The Cambridge Library Collection extends this activity to a wider range of books which are still of importance to researchers and professionals, either for the source material they contain, or as landmarks in the history of their academic discipline.

Drawing from the world-renowned collections in the Cambridge University Library and other partner libraries, and guided by the advice of experts in each subject area, Cambridge University Press is using state-of-the-art scanning machines in its own Printing House to capture the content of each book selected for inclusion. The files are processed to give a consistently clear, crisp image, and the books finished to the high quality standard for which the Press is recognised around the world. The latest print-on-demand technology ensures that the books will remain available indefinitely, and that orders for single or multiple copies can quickly be supplied.

The Cambridge Library Collection brings back to life books of enduring scholarly value (including out-of-copyright works originally issued by other publishers) across a wide range of disciplines in the humanities and social sciences and in science and technology.

# The Uses of Plants

*A Manual of Economic Botany*
*with Special Reference to Vegetable Products*
*Introduced during the Last Fifty Years*

GEORGE SIMONDS BOULGER

# CAMBRIDGE
## UNIVERSITY PRESS

University Printing House, Cambridge, CB2 8BS, United Kingdom

Cambridge University Press is part of the University of Cambridge.
It furthers the University's mission by disseminating knowledge in the pursuit of
education, learning and research at the highest international levels of excellence.

www.cambridge.org
Information on this title: www.cambridge.org/9781108068703

© in this compilation Cambridge University Press 2014

This edition first published 1889
This digitally printed version 2014

ISBN 978-1-108-06870-3 Paperback

# THE USES OF PLANTS.

# THE USES OF PLANTS:

## A Manual of Economic Botany,

WITH

*SPECIAL REFERENCE TO VEGETABLE PRODUCTS INTRODUCED DURING THE LAST FIFTY YEARS.*

BY

## G. S. BOULGER, F.L.S., F.G.S.,

PROFESSOR OF BOTANY AT THE CITY OF LONDON COLLEGE,
AUTHOR OF 'FAMILIAR TREES,' ETC.

LONDON:
ROPER & DROWLEY, 11, LUDGATE HILL.
1889.

To

MY BEST FRIEND AND FELLOW-WORKER,

MY WIFE.

# CONTENTS.

# THE USES OF PLANTS.

## INTRODUCTION.

### I.—ECONOMIC BOTANY MORE THAN FIFTY YEARS AGO.

'As plants convert the minerals into food for animals,' says Emerson, 'so each man converts some raw material in Nature to human use. . . . . Justice has already been done to steam, to iron, to wood, to coal, to loadstone, to iodine, to corn, and cotton; but how few materials are yet used by our arts!'* It is no mean boast for botanical science that, from the first writings of the herbalists of the sixteenth century down to the present day, her chief votaries have never allowed the charm of pure science to divert their attention from the practical application of their studies to the wants of their fellow-men. From the days when Gerard (1545-1612) sent collectors to the Levant to supply his physic-garden in Holborn, and from the foundation of our great national collection by Sir Hans Sloane (1660-1753), whose correspondents

* 'Representative Men,' London, 1870, p. 4.

sent him specimens from almost every corner of the
globe then known, down to our own time, British
naval and commercial enterprise, and that love of
travel for its own sake that forms one of our most
marked national characteristics, have been adding
to our knowledge of the uses of plants. Our old-
established Botanical Gardens have long carefully
collected those species that have been used in medicine. The Physic Garden of Oxford, founded by the
Earl of Danby in 1632, was the first. In 1690 the
gardens at Hampton Court were placed by William
and Mary under the charge of the botanist Plukenet
(1642-1706), who sent collectors abroad ; and about
the same date Sloane presented the garden at Chelsea,
afterwards rendered famous by the encyclopædic
works of Philip Miller (1691-1771), to the Society of
Apothecaries. In 1760 the Botanic Garden at Kew
was established by the Princess of Wales, under the
advice of Lord Bute (1713-1792), who was an en-
thusiastic botanist ; and some notion of the assiduity
with which plants were gathered together by special
collectors during the reign of George III. (mainly
through the efforts of Sir Joseph Banks, 1743-1820)
may be formed from the fact that the second edition
of the 'Hortus Kewensis,' published in 1813, enumerates 9,800 species, as against 5,500 in the edition
of 1789. Cook's voyages, on one of which he was accompanied by Banks, the discovery of the new world
of Australasia, and its further exploration by Flinders,
Robert Brown, and Ferdinand Bauer, added enormously
to our knowledge of the plants of the world. Bauer,
the artist, died in 1826 ; but Brown survived until

1858. He brought home 4,000 species as the result of the expedition, and his ' Prodromus Floræ Novæ Hollandiæ,' completed in 1830, was thus no less important as a merely descriptive work than it was as introducing into England the Natural System of classification.

The acquisition of Linnæus's collections in 1784 by Dr. (afterwards Sir) J. E. Smith (1759-1828), and the foundation, four years later, of the Linnean Society, gave considerable impetus to botanical science in this country ; whilst the Horticultural Society, founded in 1810, under the presidency of the eminent physiologist, Thomas Andrew Knight (1759-1838), carried out Sir Joseph Banks' policy of sending out collectors. By them the ill-fated David Douglas (1798-1834) was sent to supplement the work of Kalm, of Fraser, of Nuttall, and of the Bartrams in North America ; whilst George Don was sent to Brazil, the West Indies, and Sierra Leone.

It would be difficult to exaggerate the great services rendered to botany by the successive officers of the East India Company, especially the surgeons, many of them trained in Edinburgh, who advanced the empire of Flora almost as rapidly as Clive and his successors did that of temporal sovereignty. So the work of Roxburgh, Colebrooke, and Roscoe was carried on by Wallich, Griffith, Wight, Royle, and Horsfield, and the vast collections were accumulated that, in 1880, were transferred to Kew.

Though, after the deaths of King George III. and Sir Joseph Banks in 1820, the Botanic Gardens at Kew retrograded from want of either Royal or

scientific encouragement, in spite of the exertions of
Mr. John Smith, the Curator ; yet, even then, innumer-
able travellers in every corner of the globe were adding rapidly to our knowledge.

Many plants have been so long in cultivation, and
have been so altered thereby, that we no longer know
what was their wild original form.  Such is the case
with Wheat (*Triticum vulgare*, L.), Oats (*Avena sativa*,
L.), and the Barleys (*Hordeum*).  The cultivation of
other food-plants, such as Peas (*Pisum sativum*, L.),
Beans (*Faba vulgaris*, Moench.), Plums (*Prunus
domestica*, L.), and Apples (*Pyrus Malus*, L.), in this
country ; and of the Fig (*Ficus Carica*, L.), the Grape
(*Vitis vinifera*, L.) and the Indian Corn or Maize (*Zea
Mays*, L.), in other countries, may fairly be said to be
of prehistoric antiquity.  So, too, with some other
plants not used for food, such as Flax (*Linum usitatissimum*, L.), and Hemp (*Cannabis sativa*, L.), in the
Old World, Tobacco (*Nicotiana rustica*, L.), in the
New World, and Cotton (*Gossypium*) in both hemi-
spheres.  Others, again, though generally employed in
the industrial arts for ages, can hardly be said to have
been cultivated till a comparatively recent period, if
at all.  Such is the case of most timber-trees until
within the last two centuries, and of such tropical food-
plants as the Date Palm (*Phœnix dactylifera*, L.), and
the Banana (*Musa sapientum*, L.).

The Indian expedition of Alexander the Great
and the wide-reaching conquests of the Roman
generals no doubt introduced many useful plants
into Europe.  Our cornfields still bear witness, in
their weeds, to the wide area from which the Roman

granaries were supplied.* Many plants known to the ancients were, however, lost sight of during mediæval times, though to the scientific ardour of the Moors and the careful leech-craft of the monastic orders we probably owe our knowledge of many new uses of plants. But these accretions to our powers of utilizing natural products were vastly eclipsed by the novelties that resulted from the voyages of discovery during the fifteenth, sixteenth and seventeenth centuries.

Even after these, a complete enumeration of all the plants economically applied in England, or in Europe generally, at the beginning of the present century, would represent but a small proportion of those now more or less in use, to so great an extent has the vegetable world been ransacked in search of new raw materials during the last half-century. After fifty years of a reign of prosperity, it is natural to look back and contrast our knowledge at the beginning of that period with that of the present day.

II.—THE PROGRESS OF ECONOMIC BOTANY IN ENGLAND DURING THE LAST FIFTY YEARS (1837-1887).

The year 1837 cannot be said to mark a very definite epoch in the history of this steadily continuous progress. The year before, A. B. Lambert had completed the publication of his great work on ' Pinus,' in which David Don (1800-1841) incorporated much of the researches of Coulter ; whilst Darwin, returning in

* See a paper by the present writer on ' The Influence of Man upon the Flora of Essex,' Trans. Essex Field Club, vol. iv., p. 31. 1884.

the *Beagle*, brought back the well-known and ornamental *Berberis* that bears his name. Allan Cunning-
ham had gone out to succeed his murdered brother
Richard at Sydney, where he, too, died in 1839.
John Gibson was collecting Dendrobiums in India—a
collection which, with the subsequent labours of the
Messrs. Veitch, their hybridizer Dominy, and other
collectors, such as Low, Skinner, Rucker and Bate-
man, has combined to render the cultivation of these
beautiful exotics one of the most prominent features
of the gardening of our time. Within the next few
years John Miers was to return from Brazil with the
large collections which he spent so many years in
describing ; Dr. George Gardner was to leave the same
country with 7,000 specimens, and Hugh Cuming
was to return from the Philippine Islands with his
vast collection of 13,000. Loudon was but beginning
to issue those monuments of his industry, of which
the 'Arboretum Britannicum' (1833-38) is, perhaps,
the best, and Lindley had his best work yet to do.
Giles Munby did not start for Algeria until 1839, nor
Robert Fortune for China until 1843.

In 1838 it was even proposed to abolish the gardens
at Kew ; but in 1840, after their condition had been
reported upon by Dr. Lindley, her Majesty was
pleased to transfer them to what is now the Department of Works ; and shortly afterwards Sir W. J.
Hooker, then Professor of Botany at Glasgow, was
appointed Director. Among the most important
acts of this most energetic administrator was the
establishment, in 1847, of the first public Museum of
Economic Botany in the world. The nucleus of this

museum was Sir W. Hooker's private collection, formed at Glasgow. 'No sooner was the establishment and aim of the museum generally made known than contributions to it poured in from all quarters of the globe, until, in a few years, the ten rooms of the building, with its passages and corners, were absolutely crammed with specimens. Its appreciation by the public being thus demonstrated, application was made to Parliament for a grant to defray the expense of an additional building.'* The economic collection, mainly from the British Empire—in forming which Sir W. J. Hooker was much assisted by Professor J. S. Henslow —is now in three buildings at Kew, one of which is exclusively devoted to timber.

The blow received by British agriculture in 1845, on the appearance and rapid spread of the Potato disease, undoubtedly directed attention during the next few years to new plants as articles for food or for cultivation. Though New Zealand Flax (*Phormium tenax*, Willd.), failed as a crop in Ireland, the increased importation of Maize (*Zea Mays*, L.), as an article of food, from 1846, and its partly successful introduction at least as a fodder-crop, together with the far more satisfactory introduction of Italian Rye-grass (*Lolium italicum*), especially for sewage farming, may be partly attributed to this cause. But, in spite of the stimulus of a rapidly increasing foreign competition since the repeal of the Corn Laws in 1845, it must be admitted that British agriculture during the last half-century has shown very little of that elasticity

* 'Official Guide to the Museums of Economic Botany,' p. 4. 1883.

that has distinguished our importing and manufacturing industries. Exceptions to this statement must be admitted in justice to the successful cultivation of the true medicinal Rhubarb (*Rheum officinale*, Baill.), first introduced by the illustrious Daniel Hanbury, and of the Opium Poppy (*Papaver somniferum*, L.), by the Messrs. Usher, at Bodicott, near Banbury, and to that of the Sugar Beet (*Beta maritima*, Willd., var. *altissima*) at Lavenham in Suffolk.

The greatly increased variety of raw materials now imported may, undoubtedly, be in part attributed to the adoption of Free Trade, and in part, also, to the series of International Exhibitions which were, with so much foresight, originated by the Prince Consort, and which have been so materially promoted by the interest evinced by his Royal Highness the Prince of Wales.

From the Exhibitions of 1851, 1862, and 1886, and from those at Paris in 1855 and 1867, large additions were made to the Kew collection ; and in 1880 the entire series of specimens illustrating Economic Botany in the India Museum was transferred to Kew, where no less than 4,000 of them were selected for exhibition.

The Great Exhibition of 1851 was, moreover, the immediate cause of the appearance of the first popular work devoted to the entire field of Economic Botany—the late Professor T. C. Archer's ' Popular Economic Botany,' published in 1853, a very useful little work, from which, even at this distant date, the present writer has gathered much useful information.

The discovery of the art of vulcanizing caoutchouc

in 1842, and the introduction of Gutta-percha in the
following year, practically re-created an extensive
branch of our manufactures ; whilst the spread of
education and the removal of duties, causing an
enormously expanded demand for paper, have produced a corresponding increase in the import of
both old and new materials for its preparation, chief
among the latter being Alfa (commonly called
Esparto) Grass and wood-pulp. Fibrous substances
and tanning materials have also been introduced in
great variety ; and though no remedy of vegetable
origin belonging, like quinine, to the highest grade
of importance has been discovered, many drugs of
minor value have been brought into use.

The advance, however, of those of our industries
that depend upon vegetable products has not been
the result of the introduction of new materials alone.
The rapid strides of chemistry, the employment of
steam as a motive-power, the great general improvement in machine-construction and our increased botanical knowledge of the sources of our economic
products, especially of materia medica, leading to
their cultivation in new lands, have had even greater
influence. The introduction of the Tea, Chinchona
and Caoutchouc-yielding plants into various parts of
India has had, or is destined to have, an influence
upon British trade at least equal to that of any single
fresh importation into our own country.

'Vegetable products,' writes Professor Archer, in
1853,* 'constitute nine-twelfths of the whole commerce in raw produce which employs the vast

* 'Popular Economic Botany,' p. vi.

mercantile marine of this great kingdom. They furnish us with the bulk of our food and clothing, our medicine and our building materials, and with many other necessaries and luxuries. It may be argued that most of the benefits we derive from the vegetable kingdom have been discovered without the aid of science. True, but is not this a great and powerful argument in favour of the application of scientific investigation in this department? For, if so much has been done without its aid, how much may we not hope will be effected when the principles of scientific research, which have effected such miracles in every other department, are brought to bear upon that of Economic Botany ?'

Since this was written, much has in fact been done. Government botanists in our various colonies, our consular agents and many manufacturing firms and skilled specialists at home, always ably and willingly assisted by the officials at Kew, have carried out thorough, if not always systematic, investigations into the possible utilization of what are at present waste or undeveloped substances. The literature of Vegetable Technology has accordingly become so extensive as to have had a valuable bibliographical work (Mr. B. D. Jackson's 'Vegetable Technology,' Index Society, London, 1882) devoted to it exclusively.

It is unnecessary to enumerate here *in extenso* the many hundred works to which reference has been made in the preparation of this manual. Acknowledgment must, however, be made of the writer's great indebtedness to the following books of a more general character :

'Annual Statements of Trade,' issued by the Board of Trade.

Archer, T. C. : ' Popular Economic Botany.' London, 1853.

Bentley and Trimen : 'Medicinal Plants.' 1875-79.

Bevan, G. P. : 'British Manufacturing Industries.' 16 vols. London, 1877, etc.

Christy, T. : 'New Commercial Plants.' Parts 1-10. 1878-87.

Church, Prof. A. H. : ' Food.' London, 1876.

'Encyclopædia Britannica.' Ninth edition. Vols. I.-XXII. (A—S.)

Flückiger and Hanbury : ' Pharmacographia.' London, 1874.

' Guide to the Museums of Economic Botany.' London and Kew. Part 1. Dicotyledons and Gymnosperms. 1883.

Jackson, B. B. : 'Vegetable Technology.' London, 1882.

' Journal of the Society of Arts.' 1852-87.

' Pharmaceutical Journal.' 1842-87.

Smith, John : ' Dictionary of Useful Plants.' London, 1882.

Spon, E. : ' Encyclopædia of the Industrial Arts.' 2 vols. Edited by G. André and C. G. W. Lock. 1879-82.

The question of classification is a difficult one. The merely alphabetical arrangement of a dictionary, though most convenient for reference, has the disadvantage, in so wide a subject as the present, of placing in juxtaposition things entirely incongruous. In a scientific collection or treatise covering the whole field of Economic Botany, such as the Kew Museum and the guides thereto, the Natural system of botanical arrangement is undoubtedly preferable. It is, in fact, a remarkable confirmation of the truly natural character of that system, based, as it is, upon structure, especially that of flowers, fruits, and seeds—a proof that it successfully approximates to a pedigree of the Vegetable Kingdom—that it is found to correspond in its various divisions with marked distinctions in the economic substances which the plants produce. The acrid character of the *Ranunculaceæ;* the pungent but wholesome *Cruciferæ;* the aromatic pot-herbs of the *Labiatæ;* the tough 'liber,' or inner bark of the *Urticaceæ, Thymelaceæ, Tiliaceæ,* and *Malvaceæ;* the rubber-yielding 'latex' of the *Euphorbiaceæ;* the tonic bitter of the *Cinchoneæ, Gentianaceæ,* and *Simarubeæ;* the abundant resin in the *Coniferæ;* and the poisonously narcotic character of so many of the *Solanaceæ,* all alike serve to indicate this correspondence. In a short practical manual, however, such as the present, there is a great advantage in a grouping mainly technological. Following out such an arrangement, in an order as nearly logical as it seems any linear arrangement can be, the main divisions of the subject will be : I. Foods, food-stuffs, and food-adjuncts; II. Materia medica, or medicines, drugs, poisons, and other chemical substances of vegetable origin ; III. Oil-yielding seeds, vegetable oils and materials used in the manufacture of soap and perfumery ; IV. Gums, resins, oleo-resins, and inspissated saps ; V. Dyeing and tanning materials ; VI. Fibres and paper-making materials ; VII. Wood, whether for timber, furniture, engraving or other purposes; VIII. Fodder and other agricultural products ; IX. Miscellaneous substances not belonging to either of these classes.

The Systematic Synopsis will render this grouping as simple to the botanist as to the technologist.

## PART I.—FOODS, FOOD-STUFFS, AND FOOD-ADJUNCTS.

IN no class of our wants is our dependence on the vegetable kingdom so strikingly seen as in our food. With the exceptions of water and salt, all our food is, either directly or indirectly, of vegetable origin.

There is hardly any class of plants, or any part of the plant, that has not contributed in some form to our food supply. Ferns, mosses, and club-mosses afford little or no nutrient matters; but, not to speak of candied violets and rose-petals, the fleshy corollas of the Mahwa (*Bassia latifolia*, Roxb.) form an important article of diet to men and animals in India, and have been imported into this country for the latter ; whilst in the genus *Typha*, of which the common Reed-maces are well known British species, even the pollen has been employed, both in Scinde and in New Zealand, as a bread-stuff.\* The stigmas of the Saffron Crocus (*Crocus sativus*, Bert.), the cultivation of which in England is extinct, and the unopened flower-buds of the Clove (*Caryophyllus aromaticus*, L.) and the Caper (*Capparis spinosa*, L.), the fleshy peduncles of the Fig (*Ficus Carica*, L.) or the Cashew-nut (*Anacardium occidentale*, L.), and the succulent bases of the bracts in the in-

\* Lindley, Society of Arts Lectures on the Exhibition of 1851, p. 220.

flôrescence of the Globe Artichoke (*Cynara Scolymus*, L.), are other instances of the employment of unusual structures as food.

In this department of Vegetable Technology the most striking feature in the progress during her Majesty's reign is, perhaps, the results of the vastly increased facilities of transport, steam-navigation having rendered distant lands tributary, even of their fresh produce, to our population, which has so long outgrown its home supplies of the necessaries of life. In this way many perishable articles have now found their way even into the market of our streets; whilst many improved processes have added largely to our supplies of preserved fruits and vegetables. In 1882, for instance, we imported in all 64,240,749 cwt. of Wheat, of which 24,500,000 cwt. came from the United States, 11,000,000 from India, 3,000,000 from Canada, and more than that amount from Russia, besides 52,000,000 cwt. of other corn. Owing to the continuous reduction of virgin land into cultivation in the Far West of Canada and the United States, the exporting capacities of these countries are rapidly increasing, whilst increased railway communication so facilitates this export that it is stated that Wheat, which is said to cost 40s. a quarter to grow in England, may possibly be delivered in Liverpool in a very few years at 23s. We are at present importing Potatoes most largely from France; but of raw fruits our supplies are drawn from Malta, the Azores, Canaries, Madeira, the United States, and even the West Indian islands. Thus in 1878 the Azores alone shipped over 410,000 boxes of Oranges, each containing 400, to

England; and in 1882 we imported in all over 4,250,000 bushels of Oranges and Lemons. In 1886 we imported 3,261,460 bushels of raw Apples, of which 1,647,052 came from the United States, whilst of the raw fruit ' unenumerated ' imported during the same year, 906,000 bushels came from Spain.

In 1837 Pine-apples were almost exclusively home-grown, and, necessarily, expensive ; but, writing in 1854, the late Professor Archer says :* ' The importation of Pine-apples from the Bahamas has now become an extensive trade ; more than 200,000 were imported in 1851 . . . . For export to Europe it is gathered before it is quite ripe, and usually reaches England in pretty good condition. The English-grown Pine-apple usually is from ten to twelve shillings per pound, while the imported ones rarely exceed half-a-crown for the whole fruit.' In 1867 this fruit was first exported from St. Michael's, in the Azores, 427 fruit being despatched in that year. Much capital was then invested, and in 1875-76 over 34,500 were sent out. Since then the trade has so increased that those islands, with the Madeiras and Canaries, now afford us our chief supply, though, by the use of refrigerators, they are now brought from the West Indies. Fruits of various kinds, especially Peaches and Apricots, having been exported in syrup from the United States, mainly since 1867, the preservation of the Pine-apple in this manner was first carried out at Nassau (Bahamas) in 1874.

Similarly, the trade in Bananas, of which fruit only 564 bunches were imported from St. Michael's in

* ' Popular Economic Botany,' p. 47.

1879, has enormously increased, though nothing more has been heard of the ' Plantado passado,' or dried plantain-fruit, recommended by Mr. P. L. Simmonds in 1854.* The Plantain (*Musa paradisiaca*, L.) and the Banana (*M. sapientum*, L.) are generally considered distinct species, but vary so as to defy discrimination. They produce far more food, in proportion to the space they occupy, than any other plant. One Banana plant will yield three bunches, each weighing 44 lb., in the year, which is equal to 140,000 to 400,000 lb. on three acres. Thus the same area that yields 33 lb. of Wheat, or 99 lb. of Potatoes, will produce 4,400 lb. of these fruits.† Mr. Simmonds calculated that the almost imperishable food-substance referred to could be sold in London at threepence a pound. Professor Church‡ speaks of the Banana as ' a nutritious food, having less water and more nitrogenous matter than is commonly found in fresh fruits. It contains, when ripe, much sugar, but very little starch.'

As in many other departments of the present inquiry, the progress made in the half-century can be to a great extent gauged by a comparison of the articles now in commercial use with those reported upon at the time of the Great Exhibition of 1851. Undoubtedly the greatest change that has taken place in our food-supply, so far as derived from the plant-world, during this period, is the enormously increased consumption of Maize, and the introduction

* ' Commercial Products of the Vegetable Kingdom,' 1854.
† J. Smith, ' Domestic Botany,' p. 174.
‡ ' Food,' 1876, p. 123.

of glucose, which is largely prepared from it. No doubt our consumption of many fresh vegetables and fruits, of Asparagus, Tomatoes, Spinach, Artichokes, and Mushrooms, has increased more than proportionally to the increase of population; but this has been mainly met by more extensive home market-gardening, and has not seriously affected our imports. The extension of the cultivation of Maize throughout Asia, Africa, America, and Southern Europe has been mainly the work of the last 150 years. It is cultivated with less labour, probably, than any other cereal. Prior to the potato famine of 1846 Maize was not a regular article of British commerce. In 1847, 3,614,637 quarters were imported; in 1850, 1,286,263 quarters; in the first eight months of 1876, 27,000,000 cwt.; and in 1886, 31,000,000 cwt. Of this last amount, 16,700,000 cwt. came from the United States and 7,576,612 cwt. from Roumania. The removal of legislative restrictions has led to its employment in large quantities in malting, as well as in cattle-feeding. Many preparations of maize are also popular articles of food, such as corn-flour, oswego, and maizena. It is poorer in flesh-formers than Wheat, but richer than Rice, and it contains more oil than any cereal; but $64\frac{1}{2}$ per cent. of its composition is starch. It cannot be relied upon to ripen its grain in England; but may sometimes answer as a fodder-crop, the young stems being very rich in sugar and yielding in warmer climes from 50,000 to 80,000 lb. of green fodder per acre. In Brittany it forms a useful autumn crop on sandy soil too poor for Clover or Lucerne. Three-fourths of the Maize produced in the United States is grown within 450 miles of Spring-field, Illinois. In the Western States it is sometimes grown as the cheapest form of fuel. It was not until 1855 that glucose was prepared by treating the starch with dilute sulphuric acid, which has afterwards to be neutralised and removed as sulphate of lime. In 1881, 11,000,000 bushels of Maize were converted into glucose in the United States. In 1885 there were in Germany about fifty factories engaged in the manufacture of glucose, mainly from potato-starch, in which 10,000 tons of 'hard' sugar, with little dextrin present; 20,000 tons of 'syrup,' with much; and 1,250 tons of 'colour,' or glucose burnt to caramel, were produced. It can be produced at half the price of cane-sugar, and is mainly used in brewing and confectionery. In 1886 we imported 502,567 cwt., of which 441,374 cwt. came from Germany. Other important changes in the starch and sugar industries will be mentioned in the sequel.

In attempting a classification of the vegetable foods which we must pass under review, the best method available seems to be that used by Professor Church in arranging the Food Collection at Bethnal Green Museum, first got together in 1857, and in the handbook to the collection, published by the Science and Art Department in 1876.

Taking into account both the chemical composition of the substances and the purposes which they serve in the body, food-materials may be primarily divided into 'Nutrients' and 'Food-adjuncts.' 'It is impossible,' says Professor Church,* 'to draw a sharp

* *Op. cit.*, p. 168.

line of distinction between true nutrients and food-adjuncts. There is scarcely a single article of food which does not possess some constituents which give it flavour, perfume, or colour, but which yet cannot be considered as doing any actual work in the body. But these adjuncts, in the forms of flavouring and colouring matters, etc., make our food agreeable, stimulate a flagging appetite, aid indirectly in the digestion of the nutrients, and help to render palatable food which would otherwise be wasted. More than this: some of the food-adjuncts actually furnish —along with their characteristic flavouring, stimulating or narcotic constituents—real nutrients. Cocoa and beer are examples in point. And it has been thought that the active principles of certain food-adjuncts have some power of economizing the true nutrients by arresting the rapid changes of tissue, etc., which go on in the body.'

The nutrient food-principles Professor Church arranges in four groups; viz., (i.) Water; (ii.) Salts; (iii.) Carbon compounds, 'heat-givers' or 'force-producers,' such as starch, sugar and fat, and the substances resembling starch, gum, mucilage, pectose and cellulose; and (iv.) Nitrogen compounds, 'albuminoids,' or 'flesh-formers,' such as fibrin, albumen, and casein. The food-adjuncts also form four groups; viz., (i.) Alcohol; (ii.) Volatile or Essential Oils, as contained in condiments and spices; (iii.) Acids, such as those in most fruits; and (iv.) Alkaloids, nitrogenous crystallizable substances such as caffeine, theobromine and nicotine. We may, therefore, consider the nutrients practically under the following

seven heads, the food-adjuncts forming an eighth :
(1) Starches and the Cereals or Bread-stuffs ; (2)
Sugars ; (3) Pulse, seeds rich in nitrogen, largely as
casein ; (4) Roots and tubers, watery and mostly
amylaceous, non-green 'vegetables'; (5) Stems, leaves,
fruits and whole plants, mostly eaten green, as 'vegetables,' *i.e.*, with salt ; (6) ' Fruits,' commonly so called,
watery and saccharine, containing also acids and frequently pectin ; (7) ' Nuts,' oily fruits and seeds.

## I.—STARCHES AND BREAD-STUFFS.[*]

STARCH, having the composition $C_6H_{10}O_5 + 2H_2O$,
or perhaps $C_{36}H_{62}O_{31} + 12H_2O$, is perhaps the most
important force-producer in human food.  It occurs in
the leaves, stems and seeds of most plants that contain
chlorophyll, being one of the first products of the processes of assimilation, and being stored up in many
tubers, etc., as a reserve-material, in characteristic
granules.  It is largely used for other purposes besides
food, viz., in calico-printing, in dressing textile fabrics,
for laundry purposes, as adhesive paste, in toilet powder,
and in the manufacture of dextrine, or British gum, and
of glucose, or starch-sugar.  It was formerly separated
from the grain, mainly of Wheat, by a process of
steeping and fermentation, which removed the other
constituents of the grain ; but in 1840 Orlando Jones
patented its separation by dilute caustic potash, which
rendered possible its preparation from Rice.[†]   Heat

* F. Crace-Calvert, ' Journ. Soc. Arts,' viii (1859), 87.   J.
Paton, ' Encyclopædia Britannica,' vol. xxii.
† P. L. Simmonds, ' Commercial Products of the Vegetable
Kingdom,' 1854.

(320° F.) or dilute acid convert it into Dextrine, now used in dressing muslins and crapes and for postage-stamps and envelopes. The preparation of starch from Rice and Maize and the manufacture of Dextrine were of comparatively recent introduction in 1851 ;* Glucose, made by further fermentative change, was not, as we have seen, introduced until 1855. At present laundry starch is mainly made from Rice; dextrine from Potatoes ; and glucose from Maize and Wood, in the United States, and from Potatoes in Germany.

The chief edible starches are the Arrowroots, Tapiocas, Sagos and Maizenas, and these have been recently successfully imitated by preparations of Wheat (such as Semolina), Barley, Oat and Rice Meal.

ARROWROOT is obtained from the rhizomes of various species of *Maranta, Curcuma, Canna, Tacca,* and *Manihot*—that from the West Indies and Bermudas from *Maranta arundinacea,* L. ; that from the East Indies, from *Curcuma angustifolia,* Roxb., mainly ; that from Brazil, from *Manihot utilissima,* Pohl., known by the native names of 'Moussache' and 'Cipipa' ; and that from Otaheite, from *Tacca pinnatifida,* L. fil., Pia.'† Tous-les-mois or Touloula, from St. Kitt's, is the product of a variety of *Canna indica,* L., having the largest starch-grains of any known plant.‡ Arrow-root is now largely imported from Natal ; but it is of inferior quality, fetching only 2½d. to 3½d. per lb. in

* Prof. E. Solly, Society of Arts Lectures on the 1851 Exhibition.
† Spon's 'Cyclopædia of the Industrial Arts.' 'Encyclopædia Britannica,' vol. ii, p. 631. Archer, 'Economic Botany,' p. 84.
‡ Smith, 'Domestic Botany,' p. 174.

October, 1887, when Bermuda Arrowroot was priced
1s. to 1s. 3d.* Natal Arrowroot, the product of
*Maranta arundinacea,* was first imported in 1853,
rising to £13,336 worth in 1859, but now averaging
between £2,000 and £3,000. The Durban selling
price in 1886 was about 36s. per cwt.†

TAPIOCA has long been prepared in Brazil, Peru,
Guiana, Africa, and Penang, from the large tubers
of the Euphorbiaceous *Manihot utilissima,* Pohl, and
*M. Aipi,* Pohl; the former Bitter, the latter Sweet
Cassava. Tapioca mainly differs from Brazilian
Arrowroot in having undergone a partial conversion
into dextrine by being dried upon hot-plates, instead
of in the sun. The boiled juice of the Cassava,
known as 'Cassareep,' is used in many table sauces. ‡

SAGO, the demand for which, in the finely granu-
lated form in which the Malays send it to market,
has steadily increased, is obtained from the stems of
various Palms, especially *Metroxylon Rumphii,* Mart.,
and *M. læve,* Mart., which are cultivated in the island
of Ceram, in Borneo, and at Sarawak. Inferior kinds
are derived from the Gomuti palm (*Arenga saccharifera,* Lab.), the Kittool palm (*Caryota urens,* L.), the
Cabbage palm (*Corypha umbraculifera,* L.), from *C.
Gebanga,* Bl., *Raphia flabelliformis,* L., *Phœnix farini-
fera,* Roxb., and *Metroxylon filare,* Mart., in the East
Indies, and from *Mauritia flexuosa,* L. fil., and
*Guilielma speciosa,* Mart., in South America. The

* 'Encyclopædia Britannica,' *loc. cit.* 'Pharmac. Journ.' ser.
iii., vol. vi (1875), p. 204 ; vol. vii (1876), p. 169.
† 'Official Handbook for Natal' (Colonial and Indian Exhibition), p. 84.
‡ 'Guide to Kew Museum,' p. 148.

United Kingdom in 1884 imported 346,188 cwt., valued at £195,680, mostly *viâ* Singapore.*

Various preparations of MAIZE STARCH (*Zea Mays, L.*), under the names Corn-flour, Corn Starch, Oswego Starch or Maizena, are now extensively used for puddings and blancmange, especially for children and invalids.

SEMOLA or SEMOLINA consists of the small round or oval particles of some of the harder or 'flinty' varieties of Italian Wheat which collect in the furrows of the millstones. It also is used for puddings.

RICE, or PATENT, STARCH, made from Rice (*Oryza sativa, L.*) by the caustic alkali process already referred to, is now the chief form of starch employed for laundry purposes, and is also largely used in the muslin manufacture.

The staple food of more than four-fifths of the human race is formed of the grains or fruit of various grasses, most of which are ground into meal or flour and largely used in the form of bread. These grasses are known as 'cereals.' Of these the chief are Wheat (*Triticum vulgare, L.*), Barley (*Hordeum*), Oats (*Avena sativa, L.*), Rye (*Secale cereale, L.*), and the Millets, of which the most important is Durra (*Sorghum vulgare, P.*). With them may be classed the Buckwheat (*Fagopyrum esculentum,* Mch.), which is not a grass, and Maize and Rice, which, though grasses, are not used in the main as bread-stuffs.

WHEAT, unknown in a wild state, belongs mainly to two distinct forms—Winter Wheat (*Triticum hybernum, L.*) and Summer Wheat (*T. æstivum, L.*). Though used in matting, in the manufacture of starch, and

* 'Encyclopædia Britannica,' vol. xxi, pp. 148, 149.

especially in the form of damaged flour in dressing cotton fabrics, wheat is chiefly employed for bread-making. Though from 3 to 4 million acres in Great Britain are annually devoted to the growth of wheat, and 18 million quarters of this grain are annually grown in the United Kingdom, we have to import from 47 to 64 million cwt.* This supply is derived from Germany, Russia, the Danubian provinces, Egypt and India, but especially from Canada and the United States. We also import about 15 million cwt. of flour. Macaroni, vermicelli, Italian paste, etc., are prepared from hard, highly nitrogenous varieties of wheat. Wheat and other straw is, of course, also a valuable commodity, not only for plaiting into bonnets and hats, etc., but especially as a material for paper.

BARLEY is the grain of various species or varieties belonging to the genus *Hordeum* and cultivated from so remote an antiquity as to be, like wheat, unknown in a wild state. The chief are Winter, or Six-rowed, Barley, *H. hexastichum*, L., and Spring, or Two-rowed, Barley, *H. distichum*, L., an inferior kind known in Scotland (where it grows on poor soil or exposed situations) as Bere or Bigg. Barley is used as a bread-stuff, though not so extensively as formerly. It is also used, with the husk of the grain entirely removed, as Scotch or Pearl Barley. It is, however, mainly employed for malting or fermentation for the brewing of ale or porter and the distillation of whisky and gin. For these purposes upwards of 10 million quarters are annually grown in the United

---

* A bushel of wheat weighs 55-64 lb., so that a quarter weighs about 4 cwt.

Kingdom and from 6 to 10 million hundredweights are imported, mainly from the North of Europe.

OATS (*Avena sativa*, L.), though a grain also used from ancient times as a bread-stuff, and still employed in Scotland for oatcake and oatmeal porridge, are grown and imported (both on a very large scale) mainly as cattle food. Nearly 3 million acres in Great Britain are annually devoted to the cultivation of Oats. When decorticated and partially crushed they are known as Groats and are used for gruel for invalids.

RYE (*Secale cereale*, L.), said to be a native of the Crimea, is largely grown as a bread-stuff in Northern Europe; but not to any great extent in the British Isles. The grain is frequently blackened, distorted and 'spurred' by the fungus known as Ergot (*Claviceps purpurea*, Tulasne), the ' Secale cornutum ' of druggists.

DURRA, GUINEA CORN or TURKISH MILLET, also known in India as Joar, is the grain of various forms of the genus *Sorghum*, native in India, though now cultivated in Southern Europe and in the United States. In the West Indies it is known as Negro Corn and is made into cakes, but with us it is mainly used for feeding poultry. In the United States the Shaker communities obtain sugar from *Sorghum* and make carpet-brooms and brushes out of the dried fruit-panicles. Other millets grown in India and on the Continent of Europe are *Panicum miliaceum*, L. and *P. italicum*, L.

BUCKWHEAT (*Fagopyrum esculentum*, Mœnch), belonging to the natural order *Polygonaceæ*, probably a

3

native of Central Asia, is used in the United States for making cakes, but is imported and grown in this country as food for pheasants.

MAIZE or INDIAN CORN (*Zea Mays*, L.), a native of North America, is now largely cultivated in South Europe. Though it will ripen its fruit in England, it can only be relied on as a green fodder crop. When pearled it is called 'samp ;' when split, 'hominy ;' and when ground and boiled, ' mush ' in the United States, and 'polenta' in Italy. Our importation of maize as food for human beings, horses, poultry, etc., now reaches 2 million tons. The sheathing leaves of the cobs are used in South Europe for packing oranges.

RICE (*Oryza sativa*, L.), a native of India, is grown extensively in China, in Carolina, Central America, and even in the South of Russia, and is said to furnish food to one-third of the human race. We import it in enormous quantities, for food and for the manufacture of starch, from Bengal, especially Patna and Dacca, from Arracan and from Carolina, retaining nearly 200,000 tons, or upwards of 10 lb. per head, for home consumption. When in the husk it is known as 'paddy.'

MANNA CROUP is the 'groats,' or decorticated and partly crushed grains, of *Glyceria fluitans*, R. Br.,- a common grass, imported in small quantities from Russia, especially Poland.

Among other farinaceous substances* mention may be made of the grass known as CANADIAN RICE

* 'Pharm. Journ.,' ser. iii, vol. iii (1873), p. 833.

(*Zizania aquatica*, L.), which was unsuccessfully introduced some years ago, both as a food and as a paper-material, into the Lincolnshire fens.

2.—SUGARS.

SUGARS are colourless, soluble, crystalline solids, having the composition $C_{12}H_{22}O_{11}$ ('saccharons') or $C_6H_{12}O_6$ ('glucoses'). Sugar has never been made synthetically, though readily formed from starch or cellulose. Fahlberg's 'Saccharine,' now introduced for the use of diabetic or gouty subjects, prepared from coal-tar, is not a true sugar, nor is it in any way nutritive. It has the composition $C_6H_4\genfrac{}{}{0pt}{}{CO}{SO}NH$, and is many times sweeter than cane-sugar, but passes through the body unchanged.*

The chief sugars are Cane, Maple, Beet, Sorghum, Palm and Starch.

CANE-SUGAR is the product of the large grass *Saccharum officinarum*, L., a native of Southern Asia, growing ten or twelve feet high, and now largely cultivated in tropical and sub-tropical countries generally. Our supplies are now derived from the West Indies, Brazil and Mauritius. The canes are crushed, and crystalline 'raw' or 'brown' sugar and uncrystallized 'molasses' are obtained, subsequent refining yielding White or Loaf Sugar, Treacle and Golden Syrup. From the molasses Rum is distilled.

MAPLE-SUGAR, though its production is somewhat declining and it is almost entirely consumed in the

* 'American Chemical Journal,' i, pp. 170, 425 ; ii, 181.

country of production, is occasionally imported. It is prepared from the spring sap of *Acer saccharinum*, L., and the allied species *A. pensylvanicum*, L., *A. Negundo* L., and *A. dasycarpum*, Willd., one tree yielding about four pounds annually for many years. From 30 to 50 million pounds are made annually in the United States, especially in Vermont, New York, Ohio and Pennsylvania; and Canada made more than 20 millions in 1885.*

The cultivation of the WHITE SUGAR-BEETS (*Beta alba*, D.C.) was first encouraged by Napoleon I., during the war with England, and has since been supported by bounties. In 1880-81, 1,774,545 tons were made on the Continent of Europe, as against 1,979,900 tons of cane-sugar from the rest of the world. In 1884-85 the figures were 2,546,000 tons of Beet-sugar to 2,260,100 tons of Cane-sugar. In 1837 a Beet-sugar refinery was established at Chelsea, and the roots were much planted at Wandsworth in 1839. The Silesian variety, the best for the purpose, was successfully introduced into Berks, Bucks, Norfolk and Suffolk, about 1868-70.† The average yield is 7 lb. of sugar from 100 lb. of beetroots.

SORGHUM-SUGAR, obtained from the stems of the

---

* Reports of the U.S. Government Board of Agriculture.
† 'Beet-root Sugar,' Arnold Baruchson, London, 1870, ed. 2 ; Its Manufacture, W. Crookes, London, 1870, 8vo. ; Lewis S. Ware, Philadelphia, 1880, 8vo. ; see also A. Voelcker in 'Pharm. Journ.,' ser. iii, vol. i (1871), p. 854 ; and 'Journ. Soc. Arts,' vol. xix, No. 953, 1871 ; P. L. Simmonds, 'Commercial Products of the Vegetable Kingdom,' 1854 ; 'Encyclopædia Britannica,' vol. xxii, *sub voce* 'Sugar,' by James Paton ; C. G. Warnford Lock, G. W. Wagner, and R. H. Harland on Sugar Growing and Refining, 1882.

Guinea corn (*Sorghum saccharatum*, Willd.), or sugar-millet, has long been grown in China, and has been introduced successfully into the United States, France, Italy, South Russia and Australia. Several varieties are grown in America, of which the best, Minnesota Early Amber, was tried at Belvoir by W. Ingram in 1880, but not very successfully. It may, however, succeed as a fodder-plant. Its cultivation is declining in America, only 600,000 lb. being made ·in 1885.* The Zulus cultivate an allied species (*S. caffrorum*), under the name of 'Imphec,' both for sugar and for its starchy seeds.

PALM-SUGAR has been introduced into this country since 1830, often, it is believed, as Cane-sugar. In 1850 it was estimated to form one-fifth of the sugar imported ; but it is mainly consumed in India. It is obtained from the 'toddy' of several palms, especially *Phœnix sylvestris*, Roxb., whence it is termed 'Date-sugar.' The 'toddy' flows from incisions in the spadix, an average tree yielding about 35 lb. of 'jaggery,' or raw sugar, annually. Other species which yield 'toddy' are *Borassus flabellifer*, L., the Palmyra Palm, *Cocos nucifera*, L., the Cocoa-nut, *Saguerus saccharifera*, Bl., the Gomuti, *Caryota urens*, L., the Kittool, and *Nipa fruticans*, Thunb.

We consume in all upwards of a million tons of sugar, or nearly 70 lb. per head of our population, annually.

* W. Ingram, London, 1881, 12mo. ; F. L. Stewart, Philadelphia, 1867, 8vo. ; New York, 1880, 12mo.

### 3.—PULSE.

Of these leguminous plants, the seeds of which are rich in the nitrogenous substance known as legumin' and are known generally as pulse, the most important are Peas (*Pisum sativum, L.*), Beans (*Faba vulgaris*, Mœnch), Haricots (*Phaseolus vulgaris*, L.), French Beans (*P. multiflorus*, Willd.), and Lentils (*Lens esculenta*, Mœnch).

PEAS (*Pisum sativum*, L.), a cultivated form of prehistoric origin, probably derived from a plant native to the shores of the Black Sea, are largely eaten, both in an unripe state, as a green vegetable, and when ripe, as dried and split peas and pea-flour. Peas are rich in the albuminoid casein, containing 22 per cent. of this substance. 'For 1 part of flesh-formers in peas, there are only 2½ parts of heat-givers, reckoned as starch. One pound of peas contains flesh-formers equal to 3½ oz. of the dry nitrogenous matter of muscle or flesh. According to Frankland, 1 lb. of dry peas, when digested and oxidized in the body, might liberate force equal to 2,341 tons raised 1 foot high. The greatest amount of external work which it could enable a man to perform is 468 tons raised 1 foot high.'*

FIELD, BROAD, or WINDSOR BEANS, the seeds of *Faba vulgaris*, Mœnch (= *Vicia Faba*, L.), a native of Persia, are also eaten when unripe. Large quantities are imported, in addition to those of home-growth, as food for horses.

FRENCH, or KIDNEY BEANS are the unripe seed-

* Church, ' Food,' p. 83.

pods, or legumes, of *Phaseolus vulgaris*, L., a native
of India, cultivated in this country since the end of
the sixteenth century, and also extensively on the
Continent and in America. Its ripe seeds, known as
Haricots or Haricot Beans, are largely used for food
in France, but less so with us.

The SCARLET-RUNNER BEAN is the unripe pod of
*Phaseolus multiflorus*, Willd., believed to be a native of
Mexico, the ripe seeds of which are unwholesome.

LENTILS are richer in casein than either peas or
beans, containing 24 per cent. REVALENTA, or
Ervalenta, an invalid food, introduced about 1845,
and sold at high prices, consists mainly of the meal
of Lentils (*Lens esculenta*, Mœnch).* Lentils themselves have a small sale in England considering their
high nutritive value. The smaller ' Egyptian ' species
is used mainly as a cattle-food.

CHICK-PEAS, or Gram (*Cicer arietinum*, L.), are
occasionally imported and used, it is alleged, to
adulterate Coffee.

GROUND-NUTS, Monkey-nuts, Earth-nuts, or Pea-
nuts, the subterranean fruits of *Arachis hypogæa*, L.,
largely grown in the tropics, formerly only occasionally imported, are now regularly so. They are eaten
largely by poor children here as ' monkey-nuts,' and
in America in ' pea-nut candy.' They yield as much
as 50 per cent. of a bland oil, used in India as a
substitute for olive or for gingelly oil, and in Europe,
since 1840, in soap-making, pomades, cold cream, etc.
The oil sells at 20s. to 30s. per tun, and the kernels

* 'Pharm. Journ.,' iv (1845), 415; viii (1848), 30; x (1850),
64, 309.

at £8 to £15 per ton, in London (October, 1887). The cake is a useful cattle food. In 1874, 145 million pounds were exported from West Africa, mainly to Marseilles, London, Hamburg, and Berlin.*

### 4.—ROOTS AND TUBERS.

Of these more watery vegetables, the Potato (*Solanum tuberosum*, L.) is the most important. Its tropical substitutes, the Sweet Potato (*Ipomœa Batatas*, Lam.), and the Yams (*Dioscorea*), and the Jerusalem Artichoke (*Helianthus tuberosus*, L.), are somewhat similar in composition ; the Turnip (*Brassica campestris*, L., sub-sp. *Rapa*), Carrot (*Daucus Carota*, L.), Parsnip (*Pastinaca sativa*, L.), Beetroot (*Beta rubra*, D.C.), and Onions (*Allium Cepa*, L.), less so.

The POTATO is the tuber, or underground enlargement of a branch of *Solanum tuberosum*, L., a native of Chili, and apparently of Peru and Mexico. It was brought to Ireland by Sir John Hawkins in 1565 ; and to England by Sir Francis Drake in 1585, and a year later by Sir Walter Rawleigh ; and is engraved in Gerard's ' Herbal,' published in 1597 ; but did not become popular until late in the last century. It is now only second in importance among our vegetable foods to the cereals ; and, besides being extensively cultivated, especially in Ireland and Scotland, is imported in large quantities from France, Portugal, etc. Since 1845 our potato crops have in damp summers been devastated by a fungoid mildew (*Phytophthora infestans*, De Bary). The potato con-

* ' Encyclopædia Britannica,' vol. xi.

tains 75 per cent. of water and 15 per cent. of starch.
This starch is but little affected by the disease. It
has been already mentioned as a substitute for arrowroot and as a source of glucose. The latter substance
is obtained by boiling the starch with dilute sulphuric
acid ; and is largely used in brewing and in distilling,
yielding Potato spirit or British Brandy. From Potato-
starch also, by roasting, British gum, or Dextrin, used
for postage-stamps, etc., is also prepared. Whilst the
Potato belongs to the Nightshade Order (*Solanaceæ*),
the Sweet Potato belongs to the *Convolvulaceæ*, and
the Yams to the Monocotyledonous *Dioscoreaceæ*.

The SWEET POTATO (*Ipomæa Batatas*, Lam.),
unknown in a wild state, is extensively cultivated
throughout the tropics and in Southern Europe. It
is imported in small quantities from Spain. Its
tubers reach a far larger size than those of the Potato,
from which they differ also in containing 3 per cent.
of sugar. An allied species, *Ipomæa chrysorhiza*,
Hook. fil., is eaten under the name of Kumarah by
the Maoris of New Zealand.

YAMS are the tubers of several species of *Dioscorea*
(*D. sativa*, L., and *alata*, L., natives of India, *D. Batatas*,
native of China, etc.), which sometimes reach a weight
of 30 or 40 lb., and resemble potatoes even more
closely than the Sweet Potato. *D. Batatas*, the
Chinese Yam, yields enormous crops in France and
Algeria, and is hardy here, but is not appreciated.

JERUSALEM ARTICHOKES, the tubers of *Helianthus tuberosus*, L., a Composite plant, introduced from
the northern United States early in the seventeenth
century, but a native of Mexico or Brazil, take their

name from resembling the true Artichoke (*Cynara Scolymus*, L.) in flavour, and from the Italian 'girasole,' the equivalent of our old English name for Sunflower—'Turnsole.' They contain no starch, so do not become floury on boiling, but are rich in sugar and in a soluble substance resembling starch, known as 'inulin.'

The TURNIP (*Brassica Rapa*, L., a variety of *B. campestris*, L.) is the enlarged root of a biennial Cruciferous plant, which occurs wild in England. It contains 92 per cent. of water, and no starch or sugar, but a pungent essential oil. The Swede and the Rape are distinct varietal forms.

The CARROT (*Daucus Carota*, L.), a cultivated variety of a British plant, is said to have been introduced into cultivation from Holland during the reign of Elizabeth. The tap-roots contain 89 per cent. of water and 4·5 per cent. of sugar.

The PARSNIP (*Pastinaca sativa*, L.), like the Carrot, is the tap-root of a British species belonging to the *Umbelliferæ*. It has been cultivated since Roman times, and is generally eaten with salt meats. It contains 81 per cent. water, 3 per cent. sugar, and 3·5 per cent. starch, and is occasionally fermented into a wine or beer, or distilled for spirit.

The BEETROOT (*Beta rubra*, D.C., Nat. Order, *Chenopodiaceæ*), believed to be a variety of a common seashore weed, was introduced into cultivation about two centuries ago. Its enlarged red root is boiled and eaten cold. It contains 82 per cent. water and 10 per cent. sugar.

The ONION (*Allium Cepa*, L.), the bulbous stem of

a Liliaceous plant native to the Levant, is not more nutritious than the turnip. Its cultivation is very ancient.

The PARSNIP-CHERVIL (*Chærophyllum bulbosum*, L.), a native of France, with a small root, and the SKIRRET (*Sium Sisarum*, L.), a native of China and Japan, formerly much cultivated, are both umbelliferous. They are eaten boiled, as also is the root of the RAMPION (*Campanula Rapunculus*, L.), which is considerably used in France.

SALSAFY (*Tragopogon porrifolius*, L.) and SCOR-ZONERA (*Scorzonera hispanica*, L.) are Compositæ with edible roots.

The RADISH (*Raphanus sativus*, L.), a native of China, is a Crucifer that has been in cultivation for upwards of three centuries. There are, accordingly, several varieties, red and white, spindle-shaped and turnip-shaped. Though sometimes cooked, they are commonly eaten raw as salad.

Potatoes have become such a necessary of life in the British Isles, that ever since the ravages of the fungus *Phytophthora* efforts have been made to find a satisfactory substitute. During the last few years Mr. J. G. Baker, F.R.S., of Kew, and Messrs. Sutton, of Reading, have introduced to notice two new species of tuber-bearing Solanums and some hybrid forms. Whilst *Solanum tuberosum*, L., is a native of the interior of Chili, growing at a considerable altitude, in very dry soil, *S. Maglia*, Schlect., grows by the shore in a much moister climate; as noticed by Darwin in the voyage of the *Beagle*. Originally introduced in 1822, it was overlooked as identical with the Potato until 1881. This form is probably better adapted to our climate, and is capable of improvement and hybridization. *S. Commersoni*, Duval, is a low-level plant from Uruguay and the Argentine, which will not hybridize. This important experiment is still in progress.*

The ARRACACHA (*Arracacia esculenta*, DC.), an umbelliferous plant native to the northern Andes, is largely cultivated and eaten in Venezuela, etc., and is naturalized in Jamaica. It was introduced as a substitute for the Potato about fifty years ago, but did not succeed.†

The MUSQUASH (*Claytonia virginica*, L.), belonging to the *Portulaceæ*, was similarly recommended in 1854.‡

5.—STEMS, LEAVES, FRUITS, AND WHOLE PLANTS, EATEN AS 'VEGETABLES' OR AS SALADS.

Besides the roots and tubers which we have already enumerated there are other vegetables, mostly containing the substance known as chlorophyll, or leaf-green, and therefore known as 'green vegetables.' This group contains plants of which various parts are eaten, either the whole plant, the leaves, leaf-buds, leaf-stalks, stems, inflorescences or fruits being used.

* J. G. Baker, 'Journ. Linn. Soc.' (Botany), 1884, pp. 489-506, pl. 41-46. Christy, 'New Commercial Plants,' No. 7, p. 41 ; No. 8, p. 19.
† Smith, 'Domestic Botany,' p. 358.
‡ P. L. Simmonds, 'Common Products of the Vegetable Kingdom,' 1854.

They are mostly eaten boiled, and with salt and not with sugar, those eaten raw being known as salads. Of these the most important, perhaps, are the varieties of the Cabbage.

The CABBAGE (*Brassica oleracea*, L.), a maritime British Cruciferous plant, has been in cultivation for ages, and has given rise to several very distinct races of esculents. In the Channel Islands, by picking off the lower leaves, the stem is made to grow to a height of twelve feet or more, and is used for walking-sticks. Forms with loosely-arranged, enlarged, and fleshy leaves (*forma acephala*) are known as Bore-cole or Cow-cabbage. Those in which the leaves are in a compact rounded head, including the Red Cabbage, used for pickling, and the White Cabbage, are *formæ capitatæ*; and those in which the disc of the leaf is increased so that its surface rises into little hollow projections (*forma bullata*) are Savoys or Curled Kail. A number of small leaf-buds on the stem (*forma gemmifera*) constitutes Brussels Sprouts; and an abnormally enlarged, branched, and fleshy flower-stalk with abortive flowers (*forma Botrytis*) is the Broccoli and Cauliflower form.

SPINACH is the name applied to various plants, mostly *Chenopodiaceæ*, the leaves of which are eaten boiled. *Spinacia oleracea*, L., supposed to be a native of Western Asia, the leaves of which are rich in potassium nitrate ($KNO_3$), has been longest in cultivation. *Atriplex hortensis*, L., Mountain Spinach or Garden Orache, a native of Eastern Europe, is occasionally used, especially in France, as are also the leaves of small varieties of Beet (*Beta vulgaris*, L.). *Cheno-*

*podium auricomum*, Lindl., Australian Spinach, recently introduced, is one of the few food-plants we have received from Australia. *Tetragonia expansa*, Willd., Summer or New Zealand Spinach, belongs to a distinct order—the Tetragonieæ.

SEA-KALE (*Crambe maritima*, L.) is another British Crucifer, only cultivated within the last two centuries, and but little altered thereby. Its stems and leaf-stalks are blanched by being earthed up, and eaten boiled.

CELERY (*Apium graveolens*, L.), a British Umbelliferous plant, strongly smelling and unwholesome when green, is similarly treated. It is also eaten raw. Its consumption in London is very large.

ASPARAGUS consists of the young annual leafy shoots of the liliaceous *Asparagus officinalis*, L., a native of our sea-coast, cultivated since Roman times. It is now in enormous request, and is largely imported. The shoots or tops of Nettles (*Urtica dioica*, L., etc.) and of Hops (*Humulus Lupulus*, L.) have been similarly employed.

The ARTICHOKE (*Cynara Scolymus*, L.) is a Composite from the Mediterranean region, of which the edible portion is the common receptacle and the fleshy bases of the large imbricated bracts of the thistle-like inflorescence.

The CARDOON (*C. Cardunculus*, L.) is an allied species, in which the young leaf-stalks are blanched for eating.

The LEEK (*Allium Porrum*, L.), an ally of the onion, is eaten whole.

Salad plants are rich in salts, especially potash.

The chief are Lettuce, Watercress, Cress, Mustard, Endive, Chicory, and Sorrel. The LETTUCE (*Lactuca sativa*, L.), one of the Compositæ, is slightly narcotic (see p. 102, *infra*). WATERCRESS (*Nasturtium officinale*, L.), a native plant belonging to an order—the Cruciferæ—all of which are innocuous, is pungent, and is rich in sulphur. It is now largely cultivated in running water near the Metropolis and other large cities.

CRESS (*Lepidum sativum*, L.), probably a native of Persia, and MUSTARD (*Sinapis alba*, L.), a British species, are also members of this order. They are eaten when very young seedlings. Mustard also yields the condiment so named, from its ripe seeds.

ENDIVE (*Cichorum Endivia*, L.), one of the Compositæ, is a native of Northern China. Its leaves are eaten blanched. Those of CHICORY (*C. Intybus*, L.), an allied species, the roots of which are employed to mix with coffee, are similarly used, especially in France, as are also the blanched leaves and roots of the allied DANDELION (*Taraxacum officinale*, Wiggers). SORREL (*Rumex scutatus*, L.), one of the Polygonaceæ, a native of Southern Europe, is used as a salad, especially in France. The chief fruits eaten as 'vegetables' are those of the Tomato, the Vegetable Marrow, and the Cucumber, besides the Beans already mentioned.

The TOMATO (*Lycopersicum esculentum*, Mill.), belonging to the Solanaceæ, probably native to Mexico, is eaten ripe, either raw, cooked, or as sauce. Numerous varieties with red or yellow fruit of differing form are now grown on a large scale in England, and imported fresh or in tins from America, etc.

The VEGETABLE MARROW (*Cucurbita ovifera*, L.), probably, like the Pumpkin, a variety of the common Gourd (*C. maxima*, Duch.), is, like other Cucurbitaceous fruits, eaten unripe, but cooked.

The CUCUMBER (*Cucumis sativa*, L.) is generally eaten raw. Very young cucumbers are pickled, under the name of gherkins. Vegetable marrows and cucumbers are very largely cultivated in England.

The CHAYOTE, or Choco (*Sechium edule*, Sw.), the fruits of a Cucurbitaceous plant of Mexico and Venezuela, are occasionally sold in Covent Garden. They are about four inches long, oblong, tuberculate, and yellowish.

OKRO, Okra, Bendi-kai (Canarese and Tamil, used in Southern India), or Gombo (French colonies), the unripe fruits of *Hibiscus esculentus*, L. (*Abelmoschus esculentus*, Guill. and Perr.), may also be mentioned here. Cultivated throughout the tropics, and round the Mediterranean, it has been introduced, mainly in tins from the United States, during the last twelve years, as a vegetable. It is a tapering, 5-10-angled, loculicidal capsule, 4 to 10 inches long.*

FUNGI, in spite of Dr. Badham's 'Esculent Funguses' (1847), and many other similar works, are still most imperfectly appreciated as food in this country, barely half a dozen species coming to Covent Garden. They are undoubtedly somewhat

---

* 'Encycl. Brit.,' vol. xi, *sub voce* 'Gumbo ;' 'Pharmacographia,' p. 86 ; Archer, 'Econ. Botany,' p. 142 ; 'Kew Museum Guide,' p. 17.

difficult to discriminate, and are dangerous subjects for experiment; but the Giant Puff-ball (*Lycoperdon giganteum*, Batsch), the Edible Bolete (*Boletus edulis*, Bull.), the Orange-milk Agaric (*Lactarius deliciosus*, Fr.), the Champignon (*Marasmius oreades*, Fr.), and others, might well be systematically collected as food. The United States Board of Agriculture circulates coloured plates and cooking recipes of the commoner edible kinds.

Those that are commonly eaten in England are the Mushroom (*Agaricus campestris*, L.), the Horse Mushroom (*A. arvensis*, Schæff.), and the Truffle (*Tuber cibarium*, Sibth.). Of these, the Mushroom is carefully preserved in meadows, though it occurs sporadically wild, is forced in pits, and is imported from France ; the Horse Mushroom, collected in a wild state, and, owing to its firm flesh, largely sold in London, is coarser and more suited for ketchup; and the Truffle is mainly imported from France, the English species (*T. æstivum*, Vitt.) being inferior. The Morel (*Morchella esculenta*, Pers.) and Chantarelle (*Cantharellus cibarius*, Fr.) are less commonly eaten.

IRISH MOSS, or Carraigeen (*Chondrus crispus*, Lyngbye), the commonest edible seaweed, contains much mucilage, but its nutritive value is doubtful, Messrs. Flückiger and Hanbury speaking of it* as 'much over-estimated,' and Professor Church† as 'considerable.' It was introduced as a remedy in pulmonary complaints in 1831, and is used as cattle-food, for thickening colours in calico-printing, in sizing paper

* 'Pharmacographia,' p. 681.
† 'Food,' p. 106.

and cotton ; and, in America, for fining beer.  It is collected around Sligo, on the coast of Massachusetts,* and at Hamburg.

CEYLON MOSS, Jaffna Moss, or Fucus Amylaceus (*Sphærococcus lichenoides*, Agardh.), another of the *Florideæ*, from the coasts of Ceylon, Burma, etc., was introduced to European notice by O'Shaughnessy† between 1830 and 1840, and is now used as a demulcent food-jelly for invalids.

Other seaweeds used as food are GREEN LAVER (*Ulva Lactuca*, L., and *U. latissima*, L.), used as a sauce with roast mutton, now regularly sold in London ; PURPLE LAVER, or SLOKE (*Porphyra laciniata*, Agardh., and *P. purpurea*, Agardh.) ; DULSE (*Rhodymenia palmata*, Grev.), used in Scotland and Iceland ; TANGLE (*Laminaria digitata*, Lx., and *L. saccharina*, Lx.) ; and BALDERLOCHS (*Alaria esculenta*, Grev.).

### 6.—FRUITS.

The enormously increased importation of fresh fruit of all kinds from Spain and other countries, of oranges from the Azores and elsewhere, of pineapples, and of raw apples, has been already alluded to.  This last trade, together with that in applechips, etc., is mainly the creation of the last few years.

The APPLE (*Pyrus Malus*, L.) has been in cultivation from prehistoric times, being supposed by

* G. H. Bates, ' Pharm. Journ.,' xi (1870), p. 298.
† 'Pharmacographia,' p. 681 ;  Cooke, ' Pharm. Journ.,' i (1860), p. 504, etc.

Darwin to be derived from a wild 'crab' of the Caucasus, and is now, as might be expected, represented by many distinct races. It is eaten raw or cooked, and its juice is largely fermented into cider. It is essentially a tree of temperate latitudes, and we now import the fruit largely from North America.

The PEAR (*Pyrus communis*, L.), an allied member of the same genus of Rosaceæ, is also represented by several wild and a host of cultivated varieties. Some of the finest are grown in the Channel Islands. Its fermented juice is made into perry. The flavour of pears may be due to iso-amylic acetate, an alcoholic solution of which substance is used as 'Essence of Jargonelle Pears.'*

SERVICE-BERRIES, the fruits of *Pyrus torminalis*, Ehr., known provincially as 'chequers;' the QUINCE (*Cydonia vulgaris*, Pers.), used mainly for flavouring; and the MEDLAR (*Mespilus germanica*, L.), are fruits of allied trees less largely consumed.

The 'stone-fruits' are the 'drupes' of the sub-order *Drupaceæ* of the Rose tribe, having a stony 'endocarp' within a fleshy 'epicarp,' enclosing a 'kernel' or seed. Plums, cherries, apricots, peaches, and almonds, belong to this group; but in the latter the kernel, not the epicarp, is eaten. PLUMS in all their varieties, including Damsons, Greengages, and the dried 'prunes' (var. *Juliana*), or French plums (var. *Catharinea*), largely imported from France, Germany (var. *prunealina*), Bosnia, etc., are forms of *Prunus domestica*, L.

---

* Church, *op. cit.*, p. 115; Armstrong, 'Organic Chemistry,' p. 256.

The CHERRY (*Prunus Cerasus*, L.), also represented by many varieties, is largely cultivated in Kent. The liqueurs known as Cherry Brandy and Kirschwasser are prepared from it.

The APRICOT (*Prunus armeniaca*, L.), with us merely a dessert fruit, is, when dried, an important article of food in the East. It is now imported, both fresh and in tins, from the United States.

The PEACH (*Amygdalus persica*, L.) is now also imported from America in large quantities, being mainly ripened under glass in this country. The NECTARINE is a smooth-skinned variety (var. *lævis*) with a distinct flavour.

The GRAPE (*Vitis vinifera*, L.), probably originally native to Western Asia, is cultivated between 30° and 40° N. lat., and in Cape Colony and South Australia, for the manufacture of wine, and in colder climates, under glass, as dessert fruit. We import over 13,000,000 gallons of wine annually for home consumption, half of which comes from Spain and Portugal. 'Raisins' are the dried fruit, 'Valencias' and 'Malagas' coming from Spain, 'Sultanas,' which are naturally seedless, from Smyrna. We import about 20,000 tons of the various kinds. 'Currants' are the dried fruit of a smaller seedless form (var. *corinthiaca*), cultivated at Corinth, whence its name, Patras, Zante, Ithaca, etc., of which we import twice as great a weight as of raisins. They must not be confused with the small black, red, and white fruits of species of *Ribes*, known as 'Currants' in our gardens. We also import several hundred tons of fresh grapes annually, mostly from Lisbon.

The ORANGE (*Citrus Aurantium,* Riss.) is more largely imported than any other fruit. Introduced into Europe apparently by the Moors, we import it now from the Azores, Lisbon, Malta, Sicily, and Florida, the St. Michael's being the smallest, and the Maltese a seedless, variety. Hitherto the imported fruit has, from the slowness of transport, been gathered when unripe, so as never to reach this country in perfection. The Seville Orange, used for marmalade, is a distinct species (*C. vulgaris,* Riss.).

The LEMON (*C. medica,* var. *Limonum,* Brand.), said to have been introduced at the time of the Crusades, is also largely imported. Of Lemons and Oranges together we import over four million bushels. The SWEET LIME (*C. medica,* var. *acida,* Brand.), also an Indian variety, though now largely grown at Montserrat for the sake of its juice ; the SHADDOCK and POMALO, or 'Forbidden Fruit' (vars. of *C. decumana,* L.), and the CITRON (*C. medica,* Riss.), which comes, mainly as candied peel, from Madeira, are imported in far smaller quantities.

The DATE (*Phœnix dactylifera,* L.), the fruit of a palm, borne in clusters of 20 lb. weight or more, has for ages been the chief article of food in parts of North Africa and Arabia. It was probably the Lotos of the Lotophagi (Lotos-eaters). The trees are preserved as estates, and handed down as dowries. Animals and men in Fezzan, where many varieties are grown, live upon them for nine months of the year. The sap of the young shoots is fermented into 'arrack.' The fruit is imported in considerable quantities.

The POMEGRANATE (*Punica Granatum*, L.) is a Myrtaceous shrub, native to Western Asia, but naturalized in the Mediterranean region. Its fruit, which resembles a brown and rosy poppy-head, is imported in increasing quantities, mainly from Portugal. It consists of two tiers of red carpels with pithy partitions and a leathery rind.

The BANANA (*Musa sapientum*, L.) and the PLANTAIN (*M. paradisaica*, L.), fruits of closely allied prolific and gigantic Monocotyledonous plants, which are rather herbaceous than arboreous, are most important articles of food throughout the tropics, and are very nutritious, containing nearly 20 per cent. of sugar and pectose (the jelly-like substance common in fruits), nearly 5 per cent. of nitrogenous, or 'albuminoid' matter, and only 74 per cent. of water. Reference has already been made to the increased importation of these fruits, gathered when green, and to the proposed introduction of 'plantado passado' (see pp. 23-24).

WATER-MELONS (*Citrullus vulgaris*, Schrad.), oval dark-green Cucurbitaceous fruits, with white flesh, are imported from the Mediterranean ; whilst a considerable variety of Melons (*Cucumis Melo*, L.) are cultivated under glass in this country.

Among English-grown small fruits, the STRAW-BERRY (*Fragaria vesca*, L.) is pre-eminent. It is cultivated on an enormous scale in the north of Kent, round Rochester, for the London market, near Aberdeen, and elsewhere. In the same division of the Rosaceæ are the wild BLACKBERRY (*Rubus fruticosus*, L.), which is collected for puddings and jam; and some varieties of which are now cultivated, especially in the United States, and the RASPBERRY (*Rubus Idæus*, L.). The GOOSEBERRY (*Ribes Grossularia*, L.), the BLACK CURRANT (*R. nigrum*, L.), and the RED and WHITE CURRANTS (*R. rubrum*, L.) are allied species of the order Ribesiaceæ, which occasionally occur in a naturalized state in England. These small fruits, besides their use in puddings and jams, are ingredients, as are also the berries of the ELDER (*Sambucus nigra*, L), in British wines.

The PINEAPPLE (*Bromelia Ananas*, L.) is not truly a single fruit, but a collection of fruits on a fleshy peduncle, or flower-stalk. Originally native to Brazil, and cultivated in hothouses in England for two centuries, it is now extensively imported from the Bahamas and Azores, both fresh and in tins, so as to be retailed at a very low price.

The MULBERRY (*Morus nigra*, L.) is similarly the united product of a cluster of flowers. The leaves of the perianth become purple and juicy. The tree has been grown in England for three centuries, being originally a native of Western Asia. It belongs to the same order (Moraceæ) as does

The FIG (*Ficus Carica*, L.). In this tree—a native of the Eastern Mediterranean region — the edible portion is an enlarged peduncle, or flower-stalk, enclosing numerous minute flowers, which in the 'Green Figs' grown in this country, especially at Worthing, Sussex, are never fertilized, as they are, by insect agency or artificially, in their native countries. Hundreds of tons of figs are annually imported in a dried state, mainly from Smyrna. In this condition they contain $57\frac{1}{2}$ per cent. of glucose, $17\frac{1}{2}$ per cent. of water, 7 per cent. of cellulose, 6 per cent. of albuminoid matter, 5 per cent. of pectose and gum, and 3 per cent. of starch,[*] and are, therefore, very nutritious.

The petiole, or leaf-stalk, of the RHUBARB (*Rheum Rhaponticum*, L.), a Siberian plant, cultivated in England since 1628, and valued for its acidulous taste, due to oxalates, is, of course, not a fruit ; but is mentioned here as being eaten with sugar, and not, as are those plants usually called ' vegetables,' with salt.

CRANBERRIES (*Oxycoccus palustris*, Pers., and *O. macrocarpus*, Pers.) are now largely imported, the former from Russia, the latter mainly from the neighbourhood of Berlin, Wisconsin, *viâ* Chicago.

LITCHIS (*Nephelium Litchi*, Camb.), the fruit of a Sapindaceous tree, a native of China, cultivated in India, now forms a regular article of import, selling at 3s. to 4s. per lb., and LONGANS (*N. Longana*, Camb.) are also occasionally seen.

Among other less common fruits are the TAMARIND (*Tamarindus indica*, L.), the pulpy seeds of a Leguminous shrub, imported in a preserved state from both the East and West Indies ; the GUAVA (*Psidium Guajava*, L.), a Myrtaceous tree of Central America, imported as jelly ; the MANGO (*Mangifera indica*, L.), belonging to the *Anacardiaceæ*, eaten ripe in India, but only unripe and pickled in this country ; GRANADILLAS (*Passiflora quadrangularis*, L.) and POMMES D'OR (*P. maliformis*, L.) from the West Indies ;

* Church, *op. cit.*, p. 121.

' CHINESE PASSION - FRUITS,' some Solanaceous species ; LOQUATS, or Japanese Medlars (*Eriobotrya japonica*, Lindl.), first introduced by Robert Fortune ; KUMQUATS (*Citrus japonica*, Hook. fil.), another of Fortune's discoveries, which are sent over in syrup ; YANG-MAES (*Myrica Nagi*, Thunb.) of Chusan, a hardy plant introduced by Fortune in 1844 ; JAPONICAS (*Zizyphus sinensis*, Lam.) ; and the PRICKLY PEAR, or 'INDIAN FIG' (*Opuntia Ficus-indica*, Webb), which has become a regular article of importation from Algeria.

CAROB BEANS, St. John's Bread, or Algaroba (*Ceratonia Siliqua*, L.), pods containing a quantity of saccharine pulp, but ' by no means deficient in flesh-formers,'\* were largely used for our cavalry-horses in the Peninsular War, and are now exten-sively imported for the manufacture of cattle-food, though in 1854 they were but little known in this country.† They are eaten by children, but contain butyric acid, which is apt to become rancid, and being hygroscopic, they are also liable to mouldiness.

### 7.—NUTS.‡

Nuts are rich in nitrogenous matter, in addition to the 25 to 50 of fixed oil or fat.

The HAZEL-NUT (*Corylus Avellana*, L.) is the most familiar. Its varieties, the Filberts, white and red (vars. *alba* and *rubra*), and the Cob-nuts (vars. *grandis*,

\* Church, ' Food,' p. 124. See also ' Kew Museum Guide,' p. 50.
† P. L. Simmonds, 'Commercial Products of Vegetable Kingdom,' 1854.
‡ P. L. Simmonds, in ' Journ. Soc. Arts,' xx (1872), p. 475, and ' Pharm. Journ.,'ii (1872), pp. 958 and 1037.

*glomerata*, and *crispa*) are largely grown in Mid-Kent. The Filberts are characterized by the long husk, which entirely encloses the nut. Many thousand bushels are annually imported as 'Barcelona nuts' from Tarragona. They belong to the varieties *Barcelonensis, ovata, pumila,* and *variegata.* TURKEY NUTS (*C. Colurna,* L.), imported from Smyrna, are a distinct species. The oil of the nut is valuable to watchmakers and artists.

The WALNUT (*Juglans regia,* L.), native of the Himalayas, Persia, and the Caucasus, now grown throughout temperate Europe, is eaten unripe as a pickle, and is imported, mainly from Germany and the South of France, to the extent of several thousand bushels annually, partly for oil, but mainly as a dessert fruit. The 'marc,' or cake, after the expression of the oil, forms a valuable cattle-food used in North Italy. The timber of the tree is valuable, as is that of allied species, and other members of the order *Juglandeæ,* such as :

The HICKORY-NUTS (*Carya alba,* Nutt., and *C. tomentosa,* Nutt.) and PECCAN-NUTS (*C. olivæformis,* Nutt.). These American nuts are not imported in large quantities ; whilst the BUTTER-NUT (*Juglans cinerea,* Michx.) and the HOG-NUT (*Carya porcina,* Nutt.) of Canada are still more rarely seen in trade. The Peccan-nut comes mainly from New Orleans.

BRAZIL-NUTS, or Castanhas (the seeds of *Bertholletia excelsa,* H. B.) have long been largely imported from Para, not only as food, but as yielding a fine sweet oil.* About twenty-four of these seeds are

* John Miers, ' Pharm. Journ.,' v (1875), p. 726.

contained in one fruit of the enormous Myrtaceous tree by which they are produced. The allied SAPU-CAIA-NUT (*Lecythis Zabucajo*, Aubl., and *L. Ollaria*, L.) and the Camelliaceous SOMARI-NUT (*Caryocar nuci-ferum*, L., *C. tomentosum*, Willd., etc.) of Demerara, are only used here, as dessert fruit, to a far less extent.

The SWEET ALMOND (*Amygdalus communis*, L., var. *dulcis*) is a Rosaceous fruit, cultivated for its kernels in the Mediterranean area. The varieties known as Jordan (corrupted from 'Jardin,' garden, *i.e.*, cultivated) and Valencia Almonds are imported from Malaga without the shell or endocarp; the smaller Barbary or Italian sort come over in this shell. The BITTER-ALMOND is a mere variety (var. *amara*) imported from Barbary for the sake of its essential oil, which is used in confectionery, but is dangerous, owing to the presence of traces of hydrocyanic or prussic acid (HCN).

The CHESTNUT (*Castanea vulgaris*, Lam., *C. vesca*, Gaertn.), generally known as the Spanish Chestnut, a native of Western Asia, is a starchy nut, containing little oil, which forms a staple article of food in Spain and Italy. Though grown in England, it is mainly imported from these countries, Holland and Belgium.

The COCOA-NUT, or COKER-NUT (*Cocos nucifera*, L.), the fruit of one of the most useful of the tropical palms, is extensively imported. The uses of this plant are said to be as numerous as the days in a year. The timber is known as 'Porcupine-wood'; the leaves are used for thatch; the drupaceous fruits, twenty or twenty-five of which are produced by a tree every

month consist of a membranous husk or 'epicarp'
enclosing a 'mesocarp' of the valuable 'coir' fibre
used for matting, etc.; and the hard shell, or 'endocarp, used for cups; the white 'albumen' and 'milk'
serve as food and drink, and the former, on pressing,
yields an oil used for cooking, lamps and candle-
making; and the sap of the 'spadix,' or flower-head,
yields sugar, and, on fermentation, 'toddy' and
arrack.'

The PISTACHIO-NUT (*Pistacia vera*, L.), one of the
Anacardiaceæ, native of Western Asia, and much
cultivated in the Greek Islands, has a kernel green
with chlorophyll, resembling the Sweet Almond in
taste, and considerably used in French confectionery.
In the allied CASHEW-NUT (*Anacardium occidentale*,
L.) of the West Indies, not only is the roasted kernel
edible, but also the pear-like enlargement of the floral
receptacle. Though not truly a nut, mention must
here be made of the OLIVE (*Olea europea*, L.), native
to Syria and Greece, the fleshy green drupaceous
fruits of which are imported pickled in brine, but
are still more largely used for oil. Two varieties, the
large, or Spanish, and the smaller, or Italian, Olives
are imported.

DIKA-BREAD (*Irvingia Barteri*, Hook. fil.) is prepared from kernels of one of the *Simarubeæ*, growing
in profusion on the Gaboon. Containing 73 per cent.
of a solid fat, it has been suggested as a soap-making
material.

## 8.—FOOD-ADJUNCTS.

### (1) *Alcoholic.*

Alcohol is obtained from a great variety of substances by the distillation of fermented saccharine liquids. When chemically pure it is termed 'absolute,' whilst 'proof spirit' contains 49¼ per cent. of this absolute alcohol. Beer is a fermented infusion of malt, to which hops are added to make it keep and render it bitter. Malt is the germinated grain of barley, rye, wheat, maize, or rice, containing a sugar known as maltose. This is fermented by the addition of the yeast fungus (*Saccharomyces cerevisiæ*, Rees); and then the flowering catkins of the HOP (*Humulus Lupulus*, L.), the bracts of which are studded with yellow glands, containing a bitter substance known as 'lupulin,' are added. A considerable acreage in Kent, Herefordshire, and Sussex is devoted to hop-growing; but increasingly large quantities are imported, especially from Bavaria and Belgium. Much grain is imported for malting, and some yeast, in a dried state, from Germany.

Reference has already been made to the Grape-vine, Gooseberries, Currants, and Elderberries, as wine-making materials. Cowslip-flowers (*Primula veris*, L.), Oranges, Ginger, and Rhubarb are also used in 'British wines,' and artificial ethers in some foreign wines. Cider, from Apples, and Perry, from Pears, have also been mentioned.

Among distilled spirits those in common use in England are brandy, whisky, gin, and rum. Brandy

is distilled from wine, sweetened and flavoured with
prunes ; but 'British Brandy' consists largely of
'silent' (potato) spirit and caramel, or burnt sugar.
Whisky should be distilled from fermented grain,
deriving its smoky taste from traces of creosote, which
it acquires from the wood or peat smoke used in preparing it.   Gin is also—at least, when genuine—
obtained from fermented grain ; but is flavoured with
the essential oil of juniper-berries (*Juniperus communis*, L.) or with turpentine.   'Sweetened Gin' contains sugar ; cordial gin, oils of cinnamon, cloves, etc.
Rum is distilled from the molasses or uncrystallizable
liquid extract of the Sugar-cane, chiefly in the West
Indies, and is sometimes coloured with caramel or
flavoured with Pineapple.

In other countries many other substances are distilled, such as the saccharine juice of various palms,
yielding 'arrack,' and rice, yielding the Japanese 'saki.'
A great variety of flavouring substances is also used
in the sweetened spirits known as 'liqueurs' or
'cordials.'   Among these may be mentioned the
essential oil of Angelica (*Angelica Archangelica*, L.)
used in Chartreuse, and the oil of Bitter Almonds in
Noyau and Ratafia.

Under this head reference need only be made, in
conclusion, to the extensive use of Glucose, or Starch-
sugar, and of Potato, or 'silent' spirit.   The alleged
adulterants of beer are mostly far too expensive to be
used for such a purpose.   Maize is largely used in
distillation in the United States ; and, though the
' arrack ' at present obtained from the Mahwa-flowers
(*Bassia latifolia*, Roxb.) is described as a coarse spirit,

Mr. Christy has* recommended it for this purpose. It is to be regretted that, during the last few years, the sale of the very deleterious Continental liqueur ABSINTHE, manufactured from *Artemisia pontica*, L., and *Inula Helenium*, L., has much increased in our Metropolis.

## (2) *Condiments.*

These mostly owe their value, as do the Spices and Flavourers, to volatile essential oils. Some of these, such as the so-called compound ethers, have been produced synthetically.

MUSTARD is the flour or ground seed of *Brassica nigra*, Koch, and, to a less extent, of *B. alba*, Hook. fil. and Thom.—the former cultivated in Lincolnshire and Yorkshire, the latter in Essex and Cambridgeshire. It is adulterated with flour coloured with turmeric.

Mustard is apparently, however, partly the produce of *Brassica juncea*, Hook. fil. and Thom., since no less than 790 tons of its seed were imported from British India into the United Kingdom, as 'Mustard-seed,' in 1872.

TURMERIC, the trade in which is declining, since, though used in curries and to adulterate mustard, it is no longer employed as a dye, is obtained in China, Madras, Bengal, and Java from the rhizomes of *Curcuma longa*, L., but in Africa from those of a species of *Canna* cultivated at Sierra Leone.

PEPPER is the fruit of *Piper nigrum*, L., a climbing plant, native to the East Indies, but widely cultivated

* 'New Commercial Plants,' No. 2, p. 11, and No. 3, p. 36.

throughout the tropics. White Pepper is prepared from the same species by rubbing off the husk or pericarp of the berry. Other species of the genus (*P. longum*, L., and *P. officinarum*, C.DC., *P. Betle*, L., and *P. methysticum*, Forst.) yield the Dutch Long Pepper, Betel Pepper, and Kava.

CAYENNE PEPPER is prepared from the pods of species of the Solanaceous genus *Capsicum*. Those of *C. fastigiatum*, Blume, are generally known as Chillies or Pod-peppers; those of the larger, *C. annuum*, L., are known as Capsicums. Cayenne-pepper is derived from both these, and perhaps other species.

CAPERS are the flower-buds of *Capparis spinosa*, L., which are imported, pickled in vinegar, from the South of Europe. The unripe fruits of the Garden Nasturtium (*Tropæolum majus*, L.) are sometimes used instead.

Among common garden-grown condiments are HORSE-RADISH, the root of *Cochlearia Armoracia*, L.; FENNEL (*Fœniculum capillaceum*, Gilib.), the leaves of which are eaten with fish, whilst the fruits are used in cordials; GARLIC (*Allium sativum*, L.), SHALLOTS (*A. ascalonicum*, Willd.), and CHIVES (*A. Schœnoprasum*, L.), the first-named of which is a bulb of stronger flavour than the Onion, while the others are milder; TARRAGON (*Artemisia Dracunculus*, L.) and PARSLEY (*Apium Petroselinum*, L.).

Our aromatic pot-herbs, the leaves of most of which are used in a dried state, are all members of the order Labiatæ. Among them are : MINT (*Mentha viridis*, L.); SAGE (*Salvia officinalis*, L.); THYME (*Thymus vulgaris*, L.); BASIL (*Ocimum Basilicum*, L.);

MARJORAM (*Origanum Marjorana*, L.) ; and SAVORY (*Satureja*, L.).

DILL (*Anethum graveolens*, L.) and CUMIN (*Cuminum Cyminum*, L.) are Umbelliferæ with aromatic fruits, used in cordials. The latter is also employed in curry-powder and as a flavouring to Dutch cheese.

ANISE, the fruits of the Umbelliferous *Pimpinella Anisum*, L., long cultivated on, and imported from, the Continent, yields an oil consisting of Anethol or Anise-camphor, $C_{10}H_{12}O$, and variable proportions of an isomer of oil of turpentine. This substance is practically identical with that obtained from FENNEL, TARRAGON or STAR-ANISE. The latter (*Illicium anisatum*, Loureiro) has been largely substituted for anise in flavouring Dutch Aniseed and Anisette de Bordeaux. In Singapore, *I. Griffithii* and *I. majus*, Hook. fil. and Thom., are used. *I. religiosum*, Sieb. and Zucc., of Japan, a similar plant, is poisonous. *I. floridanum*, Ellis, and *I. parviflorum*, Michx., are used in the United States.*

GINGER, though, in Jamaica and India, the produce of *Zingiber officinale*, Roscoe, has been recently shown to be in Siam, and perhaps in China, the rhizome of some species of *Alpinia*.†

### (3) *Flavourers.*

It is, as Prof. Church points out (*op. cit.*), impossible to draw any very distinct line between condiments and spices ; but we may, perhaps, include the latter (which are mostly dried seeds or bark, and are eaten

---

* ' Pharmacographia ;' Spon's ' Encyclop. of Industrial Arts.'
† ' Pharm. Journ.,' xviii (1886).

with articles of food containing sugar) with the flavourers. NUTMEGS are the seeds of *Myristica fragrans*, Houttuyn, a beautiful tree of the Molucca Islands, of which about 250 tons are annually imported. Other species are now used for soap-making. MACE, also used as a spice, is the fleshy 'aril,' or outer coat, of the Nutmeg, and, when fresh, is scarlet.

ALLSPICE, or PIMENTO, is the dry berry of *Pimenta officinalis*, Lindl. (*Myrtus Pimenta*, L., *Eugenia Pimenta*, DC.), a West Indian evergreen-tree, of which we import about 2,000 tons annually from Jamaica. An aromatic oil is distilled from it; and from an allied species (*P. acris*, Wight) OIL OF BAY, used in the United States in the manufacture of the perfume known as Bay Rum, is obtained.

CLOVES are the dried flower-buds of an allied Myrtaceous plant, *Eugenia caryophyllata*, Thunberg (*Caryophyllus aromaticus*, L.), the inferior dry fruits of which are imported as MOTHER CLOVES. A native of the Moluccas, this spice is now largely imported from Penang, from Zanzibar, and from the West Indies. To the order Umbelliferæ belong the Caraway, the Coriander, and the Angelica. The CARAWAY, the ripe fruit, or rather half-fruit ('mericarp') of *Carum Carui*, L., is grown in Kent, Essex, and in the North of Europe, about 1,000 tons being imported, mainly from Holland.

The CORIANDER, the whole fruit of *Coriandrum sativum*, L., is also cultivated to a small extent in Essex, but is obtained mainly from the Mediterranean and from India.

The fruits of ANGELICA (*Angelica Archangelica*, L.)

are used in Chartreuse, and the leaf-stalks are candied as a sweetmeat.

PEPPERMINT (*Mentha Piperita*, Hudson) is a Labiate, the whole plant of which is rich in aromatic essential oil. It is cultivated at Mitcham, Surrey; Wisbeach, Cambridgeshire; Market-Deeping, Lincoln; and Hitchin, Hertfordshire; but far more extensively in New York, Michigan and Ohio, U.S.A., the oil being used for cordials and sweetmeats.

CARDAMOMS are the fruits of *Elettaria Cardamomum*, Maton, of Malabar, and various other Zingiberaceous plants,* and are employed, as are GRAINS OF PARADISE (*Amomum Melegueta*, Roscoe), to give pungency to cordials and cattle-foods.

CINNAMON is the 'liber,' or inner bark, of the Lauraceous *Cinnamomum zeylanicum*, Breyne, grown mainly in Ceylon. We import over 800 tons. Allied species, chiefly *C. Cassia*, Blume, yield the bark known as CASSIA BARK, CASSIA LIGNEA or CHINESE CASSIA, and the unripe fruits called CASSIA-BUDS, largely imported from Canton.†

Besides the fresh, dried or candied peel of oranges, lemons, and other species of *Citrus*, the essential oil of lemons obtained from them, the oil of bitter almonds and the artificial fruit-essences, to all of which allusion has already been made, the most important flavourer is Vanilla.

VANILLA,‡ after being well known in the last century, disappeared from trade, and was re-intro-

---

* Hanbury, in 'Pharm. Journ.,' xiv (1855) 352, 416; and Pharmacographia,' pp. 587-9.
† 'Pharmacographia,' pp. 474-80.
‡ Augustin Desvaux, ' Pharm. Journ.,' vii (1847), p. 73.

duced comparatively lately. It is one of the most costly vegetable productions; but Professor Morren's discovery in 1837 that the flowers of the various orchids of the genus (*Vanilla planifolia*, Andrews, etc.), from the placentæ of which the odorous substance is obtained, could be artificially fertilized* has enabled them to be successfully cultivated throughout the tropics. They owe their value to 1 per cent. of vanillin, $C_{16}H_8O_6$, a feebly acid, crystalline substance, the nature of which was demonstrated in 1858† In 1874 this substance was prepared by Tiemann and Haarmann from Coniferin ($C_{16}H_{22}O_8 + 2H_2O$), a compound first observed by Hartig in 1861 in the cambium of various conifers.‡ Coniferin is now largely collected in Continental forests and manufactured into vanillin. This substance has also been recently obtained from asafœtida.§ A former substitute for it was the Sassafras-nut, Puchury or Pichurim-bean, the large cotyledons of *Ocotea Pichurim*, H. B. K. (*Nectandra Puchury*, Nees) of the Rio Negro.‖

### (4) *Alkaloids.*

Undoubtedly the greatest service which chemistry has rendered to medicine during the present century has been the extraction from many vegetable substances of those crystalline, nitrogenous bodies known as alkaloids, by which exact doses of known character can be experimented with in a state of isolation, or

* 'Annals of Natural History,' iii (1839), p. 1.
† Gobley, 'Journ. de Pharm.,' xxxiv (1858), p. 401.
‡ 'Pharmacographia,' p. 597.
§ 'Archiv d. Pharm.,' June, 1886, p. 434.
‖ Archer, 'Econ. Bot.,' p. 94 ; 'Pharmacographia,' p. 485.

administered to the patient. Many of these act powerfully on the nervous system, generally as sedatives or narcotics ; but the action of those among them which can be classed as food-adjuncts is often modified by the presence of other substances, such, for instance, as the stimulating essential oil and the tannin which are present in tea.

As has been before suggested, a botanical identification is often nearly as important as a new discovery ; and thus Fortune's identification of the Tea-plant cultivated in China as *Thea viridis,* L., was followed, at a short interval, by its extensive introduction into India, from which country we now derive much of our most costly teas. In 1882 we imported more than 211 million pounds of tea, of which nearly 54 million were from British India, over 165 million being for home consumption.

It is also a singular fact that most nations have some non-alcoholic beverage, similar to tea, depending for its value upon an alkaloid, and that these alkaloids, though derived from different—often widely different —plants and countries, are in several cases identical, and in others most closely allied. Tea, coffee, kola, guarana, maté, and holly leaves, all contain Caffeine, or Theine, as it is sometimes called, whilst Theobromine, the alkaloid of the Cacao (*Theobroma Cacao,* L.) is very nearly related.

COFFEE affords an instance of the ease with which a trade may be destroyed. Its importation from Ceylon fell from 921,506 cwt. in 1870 to 407,222 cwt. in 1881, and 310,922 cwt. in 1885, mainly owing to the attacks of the leaf-fungus *Hemileia vastatrix,*

Berk. and Broome, and now its cultivation is mainly superseded in the island by that of the tea.* Similarly a small moth, *Cemiostoma coffeellum*, Mann, has by its ravages almost entirely destroyed the plantations in Dominica. The bulk of our coffee supply now comes from Central America. *Coffea arabica*, L., has been to some extent replaced by the more robust LIBERIAN COFFEE (*Coffea liberica*, Bull and Hiern) introduced in 1876, the ' beans ' of which are imported as 'best Java.'† Another large-leaved variety has been recently discovered, under the name ' Maragogipe,' in Brazil.‡

Of Cocoa we imported in 1886 $14\frac{1}{2}$ million pounds for home consumption.

KOLA-NUTS, the seeds of *Cola acuminata*, R. Br., a native of West Tropical Africa, belonging to the *Sterculiaceæ*, the same Natural Order as the Cacao, were first brought into notice as a new source of Caffeine about 1865.§   It is the *Sterculia acuminata* of P. Beauvois (Flore d'Oware), and is known as the Guru-nut in the Western Soudan, where it is highly valued. It is used to clear and sweeten muddy water, to assist digestion and to allay hunger. In 1882 MM. Heckel and Schlagdenhauffen showed that it contained theobromine and glucose, three times as much starch as Cacao-beans, and more caffeine in a free state than

* In 1877, 1,175 lb. of tea were exported ; in 1885, over $3\frac{3}{4}$ million.

† 'Journ. Lin. Soc.' (Botany), 1876 ; ' Pharm. Journ.,' vii (1877), p. 574.  In Ceylon, G. A. Cruewell, Colombo, 1878, 8vo. In West Indies, N. A. Nicholls, London, 1881, 8vo.  Christy : ' Commercial Plants,' No. 1.

‡ Christy : ' Commercial Plants,' No. 7, p. 79.

§ ' Pharm. Journ.,' vi (1865), pp. 450 and 457, by Prof. Attfield.

the best Coffee.* Their detailed researches† showed it to contain more theobromine than Cacao, to be closely allied to other species, especially *S. cordifolia*, Cav., introduced into the West and East Indies, Sydney, and the Seychelles, etc., but quite distinct from the Bitter, or Male Kola (*Garcinia Kola*, Heckel), which does not contain caffeine. Mr. Christy has succeeded in preparing a chocolate from them.

GUARANA-BREAD is prepared from the roasted seeds of *Paullinia sorbilis*, Mart., a strong-growing Sapindaceous climber of the Amazon Valley. The seeds are pounded and rolled into sausages, from which pieces are broken and infused,‡ both as medicine and as a drink.

MATÉ, or, more properly, Yerba de maté (maté being only the name of the gourd 'tea-pot'), the leaves of *Ilex paraguayensis*, St. Hil., and other allied species, according to John Miers, has long been culti- vated in Paraguay, where it has been drunk by all classes since the beginning of the seventeenth century. It is now in use throughout South America, where it was estimated in 1855 that 15 million pounds were consumed annually. It allays hunger, has a mildly tonic and aperient action, and keeps better than tea. Mr. Christy suggests it as valuable to our working- classes.§ It is interesting that the leaves of our

* 'Comptes Rendus,' March 20th, 1882, p. 802.
† Christy, New Commercial Plants,' No. 8 (1885), p. 5, with plate.
‡ T. C. Archer, 'Paullinia sorbilis and its Products,' 'Pharm. Journ.,' v (1863), p. 135; 'Kew Museum Guide,' p. 33; Bentley and Trimen, 'Medicinal Plants,' pl. 67.
§ Christy, 'New Commercial Plants,' No. 3, pp. 15-19.

common Holly (*Ilex Aquifolium*, L.) have long been used as tea in the Black Forest.

Mention may be made here of two adulterants of coffee, neither of which contains Caffeine, or any alkaloid :

'MOCHARA CHICORY' is prepared from roasted and pulverized Figs.*

'NEGRO-COFFEE,' brought to Liverpool from the Gambia, is the seeds of *Cassia occidentalis*, L. It proves to be very valuable as a febrifuge.†

DATE COFFEE, prepared from date-stones, has also been introduced.

## PART II.—MATERIA MEDICA.

IN no division of Economic Botany has the history of the introduction of new substances been so completely recorded as in that of Materia Medica. Various works, such as Andrew Duncan's 'Catalogue of Medical Plants' (Edinburgh, 1826), George Graves and J. D. Morries' 'Hortus Medicus' (Edinburgh, 1834), and especially Stephenson and Churchill's 'Medical Botany' (London, 1831), show us the state of the study fifty years ago. The admirably thorough records in the *Pharmaceutical Journal* from 1842 to the present time, and Lindley's 'Flora Medica' of 1846, carry on the story; whilst the complete historical synopsis in Flückiger and Hanbury's 'Pharmacographia' (1874) has received adequate illustration

* Spon, ' Encyclopædia of the Industrial Arts,' p. 707.
† Christy, 'New Commercial Plants,' No. 8, p. 40.

and has been thoroughly supplemented by Messrs. Bentley and Trimen's 'Medicinal Plants,' and the successive numbers of 'New Commercial Plants' by Mr. Christy, who has himself done so much to introduce new remedies. In these publications all that is of interest to us in the work of Continental pharmacists has been fully reproduced.

As has already been remarked, the most notable change during the half-century has been the extraction from various drugs of their alkaloids, essential oils, stearoptenes, or other principles, by which means they can be more exactly dispensed. Though Gomes in 1812 obtained a crystalline extract from Cinchona, it was not till 1820 that Pelletier and Caventon separated Quinine from Cinchonine, whilst the subsequent researches of Winckler, Pasteur, Hesse, and many others, have added enormously to our practical knowledge of the various alkaloids obtainable from this plant. So, too, nearly all the alkaloids of opium, and, in fact, those of most drugs, have been extracted during the last fifty years, until at the present day their number seems inexhaustible. Since such excellent works as those of Lindley, Flückiger and Hanbury, and Bentley and Trimen, are all arranged systematically according to the Natural System of Classification, beginning with the higher Dicotyledonous Phanerogams, in the following enumeration—and it can be little more than an enumeration—the same order will be followed.

## RANUNCULACEÆ.

*Hydrastis canadensis*, L., *Yellow Puccoon*, or GOLDEN SEAL, contains Berberia and Hydrastia, and is used as a tonic, aperient, and diuretic in North America, but not much here.[*]

*Helleborus niger*, L., CHRISTMAS ROSE, in 1852 yielded Helleborin, $C_{36}H_{42}O_6$, and, in 1864, Helleborin, $C_{26}H_{44}O_{15}$, two glucosides. With us its use is almost obsolete.[†]

*Coptis Teeta*, Wallich, from the Mishmi Hills, in Assam, first described in 1836, the bitter root of which is used as a tonic by the natives, and *Coptis trifolia*, Salisb., the 'GOLD THREAD' of North America, which also contains Berberine, and is similarly used, are neither much employed here.[‡]

*Delphinium Staphisagria*, L., STAVESACRE, contains the base Delphinine, $C_{24}H_{35}NO_2$. It was known to the ancients as Herba pedicularia, its seeds, which are imported from Trieste and the South of France, being used for the destruction of lice.[§]

*Aconitum ferox*, Wallich, the 'BISH' poison of Nepal, came into notice in 1858 as a source of ACONITINE, preferable to the European *A. Napellus*, L., which is, however, undoubtedly mixed from care-

[*] Prof. Bentley, 'Pharm. Journ.,' iii (1862), pp. 540-6 ; V. van der Espt, *ib.*, iii (1873), p. 604 ; Bentley and Trimen, pl. 1 ; 'Kew Museum Guide,' p. 7.

[†] 'Pharmacographia,' pp. 2, 3 ; Bentley and Trimen, pl. 2.

[‡] Bentley and Trimen, pl. 3 ; 'Pharmacographia,' p. 3 ; 'American Journ. of Pharmac.,' 1873, p. 193 ; Wallich, 'Trans. Med. Soc. Calcutta,' viii (1836), p. 85.

[§] 'Pharmacographia,' pp. 5-7 ; Bentley and Trimen, pl. 4.

less collection with other Alpine species. *A. ferox* is now regularly imported.*

*Cimicifuga racemosa,* Elliott, 'BLACK COHOSH,' 'Bugbane,' a native of North America, has been used here, chiefly in rheumatism, as tincture, since about 1860.†

*Xanthorhiza apiifolia,* Willd., of North America, is in repute there as a tonic, both root and wood being used.‡

### MAGNOLIACEÆ.

*Drimys Winteri,* Forster, 'WINTER'S BARK,' the bark of a South American tree, discovered in 1578 by Captain Winter, the companion of Drake, and used by him as an antiscorbutic, is now used mainly in Brazil as a tonic in cases of diarrhœa.§

*Illicium anisatum,* Loureiro, 'STAR ANISE,' to which reference has already been made, seems to have come into use as a source of a flavouring oil, not used here as medicine, about 1842. Coming from China overland, it was once known as 'Anis de Siberie.'‖

The seeds of *Magnolia Yulan,* Desf., have a reputation as a febrifuge, and the bark of *Liriodendron tulipifera,* L., is used as a stimulant tonic in America.

### MENISPERMACEÆ.

The researches of Messrs. Flückiger and Hanbury have done much to clear up the origin of the drugs

* 'Pharmacog.,' p. 9 ; Bentley and Trimen, pl. 5, 6.
† Bentley, 'Pharm. Journ.,' ii (1861), p. 460.
‡ *Ibid.,* iv (1862), p. 12 ; Bentley and Trimen, pl. 9.
§ 'Pharmacographia,' pp. 17-20.
‖ Bentley and Trimen, 'Medicinal Plants,' pl. 10 ; 'Pharmacographia,' p. 21.

known as 'PAREIRA BRAVA,' and the restoration of this remedy to the British Pharmacopœia. The true 'Pareira Brava' is *Chondrodendron tomentosum*, Ruiz and Pavon, the chief substitute for it, *Cissampelos Pareira*, L., belonging to the same order. They contain Buxine, $C_{18}H_{21}NO_3$ (formerly termed Pelosine and Bibirine), and are valued as diuretics.*

*Jateorrhiza palmata*, Miers, 'CALUMBA ROOT,' a native of the forests of East Africa, the large fleshy roots of which contain several related bitter principles, is much used as a mild tonic. It is shipped from Zanzibar.†

*Anamirta Cocculus*, Wight and Arnott, 'COCCULUS INDICUS,' though now discarded from our Pharmacopœia, is still in use in India for destroying lice, and is imported to the extent of about 50,000 lb. of the dried, berry-like fruit annually, at the price of seven to nine shillings per cwt., mainly for re-exportation. The seeds contain from four to one per cent. of the poisonous picrotoxin $C_{12}H_{14}O_5$, and so stupefy fish when thrown on the water.

*Tinospora cordifolia*, Miers ($= Cocculus cordifolius$, DC.), 'GULANCHA,' has a bitter tonic root and stem, and is used in India.

### BERBERIDACEÆ.

*Berberis aristata*, DC., *B. Lycium*, Royle, and *B. asiatica*, Roxb., yield the tonic bark known in India as 'RUSOT.'‡

---

* Flückiger, 'Neues Jahrb. f. Pharm.,' xxxi (1869), p. 257 ; 'Pharm. Journ.,' xi (1870), p. 192; Hanbury, *ib.* (1873), pp. 81, 102 ; 'Pharmacog.,' pp. 25-30 ; Bentley and Trimen, pl. 11.
† 'Pharmacographia,' pp. 22-5 ; Bentley and Trimen, pl. 13.
‡ 'Indian Pharmacopœia,' p. 436 ; Bentley and Trimen, pl. 16.

*Podophyllum peltatum*, L., of Canada and the United States, contains in its rhizome the purgative resin PODOPHYLLIN, admitted to the British Pharmacopœia from that of the United States in 1864.*

*Caulophyllum thalictroides*, Michx.,† the 'BLUE COHOSH,' and *Jeffersonia diphylla*,‡ Pers., the 'TWIN-LEAF,' belong to the same order.

### SARRACENIACEÆ.

*Sarracenia purpurea*, L., 'THE INDIAN CUP,' recommended as a remedy in small-pox, is valueless.

### PAPAVERACEÆ.

*Sanguinaria canadensis*, L., the 'BLOOD-ROOT,' or PUCCOON,' of North America, is stimulant and diaphoretic.§

*Papaver somniferum*, L., the OPIUM POPPY, said to be truly wild in the islands of the Mediterranean (*P. somniferum*, var. *setigerum*, Boissier), has been cultivated from a remote antiquity for its sedative properties. The dried milky 'latex' obtained from incisions in the unripe fruit-capsules constitutes opium, the narcotic property of which depends upon morphine, $C_{17}H_{19}NO_3$, and various other related alkaloids. Six distinct kinds occur in commerce, *viz.*, (1) that of Asia Minor or Smyrna, and Turkey or Constan-

---

* Bentley, 'Pharm. Journ.,' iii (1862), p. 452 ; 'Pharmacog.,' p. 35 ; Bentley and Trimen, pl. 17.
† 'Pharm. Journ.,' iv (1862), p. 52.
‡ *Ibid.*, p. 104.
§ *Ibid.*, pp. 263-9, and G. D. Gibb, 'Pharm. Journ.,' i (1860), p. 454 ; Bentley and Trimen, pl. 20.

tinople, and (2) that of Egypt, the produce of the round-capsuled *P. somniferum,* var. *glabrum,* Boissier, and that of (3) Persia, (4) India, (5) China, and (6) Europe, the produce of the ovate-capsuled var. *album,* Boissier. We import the drug mainly from Turkey. The capsules are used whole for fomentations, and the seed (known as Maw-seed') for oil. The residue, after the extraction of the oil, is an oil-cake, valuable for cattle. The petals of our common wild poppy (*Papaver Rhœas,* L.) are used in pharmacy merely as a red colouring agent.

*Chelidonium majus,* L., the GREATER CELANDINE, a rustic remedy for warts, gives its name to, but does not enter into the composition of, a patent cure for corns.

## CRUCIFERÆ.

*Brassica nigra,* Koch, MUSTARD, which has already been mentioned as a condiment (p. 63, *supra*), is a powerful external stimulant. It contains an albuminous substance, discovered in 1839, known as 'myrosin,' and an essential oil, Allyl sulphocyanide, $C_4H_5NS$, which also occurs in *Reseda lutea,* L., and *R. Luteola,* L. Myrosin occurs in *Brassica alba,* Hook. fil. and Thom., together with Sinalbin, $C_{30}H_{44}N_2S_2O_{16}$.*

## CANELLACEÆ.

*Canella alba,* Murray, a West Indian tree, yields an aromatic bark known in the Bahamas, whence it is imported, as 'WHITE WOOD BARK' or 'CINNAMON BARK.'†

* Bentley and Trimen, pl. 22, 23.
† *Ibid.,* pl. 26.

## BIXINEÆ.

*Gynocardia odorata*, R. Br., of India, contains in its seeds an oil known as CHAULMUGRA OIL, introduced from the Indian Pharmacopœia for rheumatism, syphilis, etc., about 1872, since when it has increased in popularity.*

## POLYGALACEÆ.

*Krameria Ixina*, L. (=*K. tomentosa*, St. Hil.), 'SAVANILLA' or 'NEW GRENADA RHATANY,' was found in 1864 to be equal, if not superior, to that from Peru (*K. triandra*, Ruiz and Pavon) as an astringent in dysentery.†

*Polygala Senega*, L., SENEGA ROOT, is a North American stimulating expectorant and diuretic.

## GUTTIFERÆ.

*Garcinia Hanburii*, Hook. fil., the main source of GAMBOGE in Cambodia, was described by Christison in 1851; but more fully, with a figure, in 1864,‡ by Hanbury, to whose memory it was dedicated by Sir J. D. Hooker in 1875.§ Gamboge consists of a resin, with fifteen or twenty per cent. of gum,‖ and acts as a drastic purgative.

* 'Gynocardia odorata,' by Richard C. Lepage, London (1878), 8vo.; Christy, 'New Commercial Plants,' No. 2, p. 3; 'Pharmacographia,' p. 70; Bentley and Trimen, pl. 28.

† Bentley and Trimen, pl. 31; Hanbury, 'Pharm. Journ.,' vi (1865), p. 460

‡ 'Trans. Linn. Soc.,' xxiv (1864), p. 487, tab. 50, under the name *G. Morella*, var. *pedicellata*.

§ 'Journ. Linn. Soc.,' xiv (1875), p. 485.

‖ 'Pharmacographia,' pp. 77-79; Bentley and Trimen, pl. 33.

*G. indica,* Choisy, yields from its seeds KOKUM BUTTER, said to have been used to adulterate ghee, and since recommended for use in pharmacy.* It consists chiefly of stearin, so might be of use in candle-making.

*Calophyllum inophyllum,* L., one of the most valuable timber-trees of the tropics, known as 'Poon' or 'Tamanu,' has in its seeds an oil known as 'NDILO' in Fiji, or as BITTER-OIL in India, recommended for soap-making, but used locally for ringworm.†

### DIPTEROCARPACEÆ.

*Dipterocarpus alatus,* Roxb., *D. turbinatus,* Gærtn. fil., and other species, yield, on incision of the stem, GURJUN BALSAM, or WOOD-OIL, the essential oil of which has the composition of Copaiba, $C_{40}H_{32}$, for which it is used as a substitute. It is imported from Moulmein and elsewhere as 'East Indian Balsam Capivi.' Its wood is used in boat-building, and the balsam as a protective varnish against white ants.‡

*Dryobalanops aromatica,* Gærtn., yields the stearoptene known as Borneol, $C_{10}H_{18}O$, BORNEO or SUMATRA CAMPHOR, which is so eagerly bought up by the Chinese that it does not come into Europe.§

### MALVACEÆ.

*Althæa officinalis,* L., the MARSH MALLOW, a British plant, has long been used as a demulcent.

* Pereira, 'Pharm. Journ.,' xi (1852), p. 65 ; 'Comptes Rendus,' xliv (1857), p. 1355 ; Bentley and Trimen, pl. 32.
† 'Pharm. Journ.,' vol. xvii (1886).
‡ Smith, 'Domestic Botany,' p. 477 ; 'Kew Museum Guide,' p. 17 ; 'Pharmacographia,' p. 81.
§ Sir W. J. Hooker, in 'Pharm. Journ.,' xii (1852), p. 300 ; 'Pharmacog.,' p. 464 ; W. T. Thiselton Dyer, Journ. Bot., 1874, p. 98.

Its root is an ingredient in the French Guimauve cough-lozenges.

Passing over the slight medical character of Kola (*vide supra*, p. 70) and the mere vehicle Cocoa-butter, we come to the

## LINACEÆ.

*Linum usitatissimum*, L., the FLAX, cultivated in Egypt and Europe from prehistoric times as a fibre-plant, yields also oily seeds known as LINSEED, which are used not only as a source of oil for lamps, and painting, etc., but also in medicine, either for a demulcent infusion or tea, or for poultices. The seed is mainly imported from Russia, nearly 2,500,000 quarters being imported in 1882.

*Erythroxylon Coca*, Lamarck (1786), COCA, the dried leaves of which are chewed to assist digestion of starchy food by the Indians of South America, has since 1885 been admitted into the British Pharmacopœia, having been in use since about 1870, and being strongly recommended as allaying hunger by the late Sir R. Christison.* From them has been obtained the valuable local anæsthetic COCAINE, already in universal use in dental and other surgery.†

## ZYGOPHYLLEÆ.

*Guaiacum officinale*, L., and *G. sanctum*, L., West Indian species, yield LIGNUM VITÆ, chips of which are

* 'Pharm. Journ.,' vi (1876), p. 883 ; see also P. Mantegazza, *ibid.*, i (1860), p. 506 ; Abl, *ibid.*, vii (1865), 33 ; Fournier, *ibid.*, i (1870), p. 43 ; Christy, 'New Commercial Plants,' No. 3, p. 24.
† W. Martindale, ' Coca, Cocaine, and its Salts,' 1886.

used in compound decoction of sarsaparilla, and a resin frequently prescribed for rheumatism and gout.*

## GERANIACEÆ.

*Geranium maculatum*, L., 'ALUM-ROOT,' is recommended for dysentery,† for which the now largely disused ANGOSTURA BARK (*Galipea Cusparia*, St. Hil.), which contains angosturine, is also employed. This last belongs to the

## RUTACEÆ.

*Barosma*, various species of which genus yield the BUCHU leaves, introduced in 1821, contains a camphor like that of peppermint. The leaves are a reputed diuretic.

*Ticorea febrifuga*, St. Hil., a Brazilian tree, has a bark recommended as a quinine substitute.

*Zanthoxylum fraxineum*, Willd., the TOOTH-ACHE SHRUB, has a bark recommended as provocative of saliva.‡

*Z. caribœum*, Lam., and *Z. Perrotetii*, D.C., of Cayenne, the ' BOIS PIQUANT,' or YELLOW or THORNY CAVALIER, is a powerful febrifuge, which may possibly rival the Cinchonas.  It contains $C_{12}H_{24}O$ as a crystalline solid and an alkaloid.§

*Pilocarpus pennatifolius*, Lemaire, 'Jardin Fleuriste,' iii., t. 263 (1852), and perhaps *P. Selloanus*, Engler, pro-

* 'Pharmacographia,' pp. 92-97. Bentley and Trimen, pl. 41.
† R. Bentley, ' Pharm. Journ.,' v (1863), pp. 20-25.
‡ *Ibid.*, iv (1863), pp. 399-407.
§ 'Pharm. Record,' 1886, p. 342 ; Heckel and Schlagdenhauffen, 'Repertoire de Pharmacie,' 1886 ; Christy, ' New Commercial Plants,' No. 9, p. 62.

duce the leaves known as PERNAMBUCO JABORANDI, containing pilocarpine and acting as a diaphoretic and powerful sialagogue. It was introduced into Europe in 1847, but not to England until 1874,* and has, in 1885, been admitted into the Pharmacopœia. It has been used with some success in hydrophobia.

*Ptelea trifoliata*, L., 'SHRUB TREFOIL, or 'WATER ASH,' has been recommended as a vermifuge, and for ulcers.†

*Toddalia aculeata*, Pers., an aromatic prickly bush, a native of India, yields a bark from its roots, used in India as a stimulating tonic.‡

## AURANTIACEÆ.

The peel of the Seville or Bitter Orange, the 'Bigaradier' of the French (*Citrus vulgaris*, Risso = *C. Aurantium*, var. *amara*, L. = *C. Bigaradia*, Duhamel), is an aromatic tonic; but the other products of the genus, the peel of the lemon (*C. Limonum*, Risso), the essence or essential oil of Lemon, essence of Bergamot, obtained from the unripe fruits of the Calabrian Bergamot Orange (*C. Bergamia*, var. *vulgaris*, Risso and Poiteau), essence of Neroli, prepared from the flowers, essence of petit grain, from the leaves and shoots, and essence of orange-peel from the unripe fruit of *C. vulgaris*, and, in the case of the last-mentioned, of *C. Aurantium*,Risso, and essence of Cedrat

* Baillon and Holmes, 'Pharm. Journ.,' iv (1874), p. 850 ; v (1874-5), pp. 364, 561, 569, 574, 581, 781 ; vi (1876), p. 750 ; vii (1876-7), p. 10, 496, 731, 892 ; viii (1878), p. 892, with plate at pl. 581 of vol. v ; Bentley and Trimen, pl. 48.

† Bentley, 'Pharm. Journ.,' iv (1863), pp. 494-7.

‡ 'Pharmacographia,' pp. 101-3. Bentley and Trimen, pl. 49.

from the fruit of *C. medica*, Risso, are used mainly in perfumery, or for liqueurs, or merely as flavourers in medicine.

*Ægle Marmelos*, Corr., the BAEL fruit of India, is used in diarrhœa or dysentery, attention having been directed to it in Europe about 1850.*

### SIMARUBEÆ.†

*Picræna excelsa*, Lindl. (= *Quassia excelsa*, Swartz = *Simaruba excelsa*, DC.), QUASSIA or BITTER WOOD, a native of Jamaica, is employed as a stomachic and tonic, its wood being often turned into 'Bitter-cups.' That of *Quassia amara*, L., of Surinam, is also used.

*Simaba Cedron*, Planchon, 'CEDRON,' of Panama, has bitter seeds, suggested as a quinine substitute and for subcutaneous injection in hydrophobia, and used locally for snake-bite. It was sent to Kew by Purdie and Seemann between 1843 and 1850.‡

### BURSERACEÆ.

*Boswellia Carteri*, Birdwood,§ was first recognised as the source of GUM OLIBANUM, or FRANKINCENSE, by Surgeon-Major H. J. Carter, in South-East Arabia, in 1846.‖  *B. Bhau-Dajiana*, Birdw., of Somali-land,

---

* 'Pharmacographia,' p. 116; 'Pharm. Journ.,' x (1850), p. 165; ii (1861), p. 499.  Bentley and Trimen, pl. 55.
† A. W. Bennett, 'Notes on Indian Simarubeæ.'
‡ Smith, 'Dictionary of Economic Plants,' Hooker's 'Journ. of Botany,' v (1846), 566.
§ 'Linn. Trans.,' xxvii (1871), p. 143.
‖ Journ. of Bombay branch, Royal Asiatic Soc., ii (1848), p. 380, tab. 23.  Bentley and Trimen, pl. 58.

Luban Maitee,' *B. Frereana*, Birdw., being often mixed with it.

*B. serrata*, Roxb., the 'Salai' of the Deccan, and *B. glabra*, Roxb., are used locally as ' INDIAN OLIBA-NUM,' for incense.*

*Balsamodendron Myrrha*, Nees, ' AFRICAN or TUR-KEY MYRRH,' *B. Opobalsamum*, Kunth, ' ARABIAN,' and *B. Kafal*, Kunth, ' EAST INDIAN MYRRH,' are, according to the researches of Dr. Trimen,† R. H. Parker,‡ Mr. E. M. Holmes and Professor Oliver, the sources of that other gum-resin which is more used in medicine than the last-mentioned.

*B. Mukul*, Hook. (' Journ. Botany,' 1849), is the source of ' GOOGUL ' or INDIAN BDELLIUM ; *B. Play-fairii*, Hook. fil., discovered in 1862, by Colonel Playfair, that of ' HODTHAI,' or OPAQUE BDELLIUM.

*Canarium commune*, L , is believed to be the source of the resin known as MANILLA ELEMI, admitted to the Pharmacopœia in 1867, though *Icica heptaphylla*, Aubl., ' HYAWA GUM,' *I. Caranna*, H.B.K., ''CARANA,' and other species, yield similar products known as BRAZILIAN ELEMI. Further information is still much needed as to the sources of these oleo-resins.

## MELIACEÆ.

*Swietenia senegalensis*, Desv., ' CAILCEDRA,' has a tonic bark, containing ' Calcedrine.'§

* Flückiger, 'Pharm. Journ.,' viii (1878), p. 407 ; Spon's 'Encyclop. Indus. Arts.'
† 'Pharm. Journ.,' ix (1878-9), No. 462, p. 893 ; Bentley and Trimen, 'Medic. Pl.,' i, pl. 59-60.
‡ 'Pharm. Journ.,' ix (1879), No. 475.
§ Christy, 'New Commercial Plants,' No. 10, p. 42.

*Melia indica*, Brandis (= *M. Azadirachta*, L.), the NIM or MARGOSA of India, contains an alkaloid in its bark, known as Margosine, used in India as a tonic and antiperiodic.*

*Soymida febrifuga*, Juss. (= *Swietenia febrifuga*, Willd.), the BASTARD CEDAR, or ROHUN, another Indian tree, has also an astringent tonic bark, valuable in dysentery.†

### RHAMNACEÆ.

*Rhamnus cathartica*, L., the BUCKTHORN, a British shrub, has berries which are collected, when ripe, in Hertfordshire, Buckinghamshire, and Oxfordshire, for their juice, from which a purgative syrup and the pigment known as SAP GREEN are prepared. The latter substance is, however, mainly prepared from Persian berries, the fruit of *R. infectorius*, L.‡

*R. Purshianus*, DC., yields a bark which has been introduced under the name of CASCARA SAGRADA ('Sacred Bark') as a tonic purgative.

### ANACARDIACEÆ.

The resins of the MASTICH (*Pistacia Lentiscus*, L.) and the Terebinth (*P. Terebinthus*, L.) § the latter known as CHIAN TURPENTINE, are practically obsolete in English medicine; though this latter has recently been suggested for use in cancer.

*Comocladia integrifolia*, Jacq., 'MAIDEN PLUM,' of the West Indies, has a bark said to be hypnotic.‖

* Bentley and Trimen, pl. 62.        † *Ibid.*, pl. 63.
‡ *Ibid.*, pl. 64                          § *Ibid.*, pl. 68, 69.
‖ 'Christy, 'New Commercial Plants,' No. 10, p. 106.

## CONNARACEÆ.

*Cnestis glabra,* Lam., is the ' MORT AUX RATS' of the Mascarene Islands.

## LEGUMINOSÆ.

*Ulex europæus,* L., has been found by Mr. A. W. Gerrard* to contain an alkaloid, ULEXINE, more powerful as a purgative than SPARTEINE, obtained from *Cytisus scoparius,* Link., whilst *C. Laburnum,* L., contains two poisonous substances, CYTISINE and LABURNINE, discovered in 1865 by Messrs. Husemann and Marne.

FENUGREEK, the seeds of *Trigonella Fœnumgræcum,* L., though still used in curry-powder, in cattle foods, in veterinary medicine, and, from containing Coumarin, in adulterating inferior hay, is obsolete in ordinary pharmacy.

*Astragalus gummifer,* Labil., and other species in Syria, Asia Minor, and Greece, have been shown† to be the sources of that mucilaginous transformation of medullary cells that constitutes the officinal GUM TRAGACANTH, which is largely imported from Smyrna.

*Glycyrrhiza glabra,* L., the LIQUORICE, a native of the Mediterranean area, cultivated at Mitcham, Surrey, and near Pontefract, Yorkshire, and imported to the extent of some hundreds of tons annually, from

---

* ' Pharm. Journ.,' Aug. 7, 1886.
† Boissier, 'Flora Orientalis,' ii (1872) ; ' Pharmacographia,' p. 152 ; ' Kew Guide,' p. 41 ; M. Giraud, ' Pharm. Journ.,' 1878, No. 405. Bentley and Trimen, ii, pl. 73.

Spain, Russia, Germany, France, and Turkey, contains in its root much sugar, and a substance known as Glycyrrhizin, $C_{24}H_{36}O_9$. It is used as a demulcent in cough-lozenges, as a sweetmeat, and in the preparation of tobacco.

*Arachis hypogæa,* L., already alluded to (p. 39, *supra*), was first imported to Marseilles from Goree, as an oil-seed, about 1840, and is largely used as a substitute for olive-oil.

*Abrus precatorius,* L., the root of which is an inferior substitute for liquorice, has been found to contain in its seeds, the well-known Crab's-eyes or Jequirity-seeds, an alkaloid JEQUIRITINE, stated to be antagonistic to Atropine.*   They have recently been used in ophthalmia, lupus, etc.

*Erythrina corallodendrum,* L., of Brazil, and other tropical countries, yields a bark said to be hypnotic.†

*Mucuna pruriens,* DC., COWHAGE, a tropical twining plant, has the pods densely covered with stiff brown hairs, which are a powerful mechanical irritant, and are administered, in honey or syrup, as a vermifuge.

*Physostigma venenosum,* Balf., the ORDEAL BEAN of Old Calabar, first made known in England by Dr. W. F. Daniell, about 1840, and described by Balfour in 1861, was found, in 1863, to contain an alkaloid, Physostigmine, having the composition $C_{30}H_{21}N_3O_4$, directly antagonistic to atropine, and is now much

---

* Heckel and Schlagdenhauffen, 'Der Fortschritt' (Geneva), 1887, Nos. 2, 3, 4 ; Bentley and Trimen, ii, pl. 77 ; Christy, 'New Commercial Plants,' No. 10, p. 107.

† Bentley and Trimen, p. 102 ; 'Nouv. Remèdes,' 1886, p. 418.

usẹd in ophthalmic and nervous diseases. The longer beans of *P. cylindrospermum*, Holmes, have been recently imported and found to be more valuable.*

*Pterocarpus Marsupium*, Roxb., was shown by Royle† to be the main source of KINO, which astringent was originally obtained from the African *P. erinaceus*, Poir. It is also obtained from *Butea frondosa*, Roxb., the 'DHAK,' or 'PALAS KINO,' and in Australia from *Eucalyptus rostrata*, Schlecht., *E. corymbosa*, Sm., *E. citriodora*, Hook., and other species.

*Piscidia Erythrina*, L., the WHITE DOGWOOD of Jamaica, is stated to possess hypnotic and anodyne properties in the bark of the root,‡ and the bark of the valuable timber tree *Andira inermis*, Kunth., of the West Indies, is said to be a drastic purgative.

*Andira Araroba*, Aguiar, 'ARAROBA,' or 'GOA POWDER,' has, since 1864, attracted considerable attention, and the pith is now, under the name 'CHRYSAROBIN,' admitted to the Pharmacopœia.§

*Myroxylon Toluifera*, H.B.K., is a native of Venezuela and New Granada, from the stems of which BALSAM OF TOLU is obtained. It was first fully described by Mr. John Weir, plant-collector to the

* Spon's 'Encyclop. Indust. Arts,' p. 795 ; 'Pharmacographia,' p. 167, and references there given ; Smith, 'Domestic Botany,' p. 431. Bentley and Trimen, ii, pl. 80.

† 'Pharm. Journ.,' v (1846), p. 495 ; 'Pharmacographia,' p. 170 ; Bentley and Trimen, ii, pl. 81.

‡ Dr. Halsey, 'Therapeutic Gazette,' August, 1886.

§ 'Pharm. Journ.,' v (1864), p. 345 ; E. M. Holmes, *ibid.*, v (1875), pp. 723, 729, 801, 816 ; *ibid.*, ix (1878-9), pp., 29, 755 ; *ibid.*, x (1879-80), pp. 42, 814, (Thos. Greenish) ; and pamphlet by Dr. Balmanno Squire.

Royal Horticultural Society, in 1863.* It is used in cough-lozenges.

*Myrospermum Pereiræ,* Royle, a tree of Central America, is the source of BALSAM OF PERU, now but little used. In 1861 the tree was successfully introduced into Ceylon.†

*Cæsalpinia Bonducella,* Roxb., and *C. Bonduc.,* Roxb., produce the seeds known as 'NICKER NUTS,' used in India, and recommended here, as a tonic.

*Hæmatoxylon Campechianum,* L., LOGWOOD, a Central American tree, now naturalized in Jamaica, and largely imported as a dye, has been used in diarrhœa.

*Cassia acutifolia,* Delile, and *C. angustifolia,* Vahl, shrubs native of North Tropical Africa, from Timbuktu to Nubia, and of Somâli-land, Arabia, and the Panjab respectively, are the main sources of SENNA. The leaves of the former are' known as Alexandrian or Nubian; those of the latter as Bombay or Tinnevelly Senna. In addition to oxalic, tartaric, and malic acids, Senna contains a substance described as Cathartic acid, on which its valuable purgative properties may depend.‡ *C. marilandica,* L., is the source of AMERICAN SENNA.§ The saccharine pulpy lining of the transverse septa in the pods of the Indian tree,

* · Journ. Royal Hort. Soc., May, 1864 ; 'Pharm. Journ.,' vi (1865), p. 60; Bentlev and Trimen, ii, pl. 84.
† D. Hanbury, 'Pharm. Journ.,' v (1864), pp. 241, 315 ; J. Attfield, *ibid.,* pp. 204-6, and H. Baillon, *ibid.,* iv (1873), p. 382 ; Bentley and Trimen, ii, pl. 83.
‡ 'Pharmacographia,' pp. 189-194 ; Bentley and Trimen, ii, pl. 90, 91.
§ 'Pharm. Journ.,' v (1846), p. 345 ; 'Kew Museum Guide,' p. 49 ; Bentley and Trimen, ii, pl. 88.

*C. Fistula*, L., now largely grown in the West Indies, is the mild laxative PURGING CASSIA.*

*Cassia occidentalis*, L., 'Café Nigre,' or NEGRO-COFFEE, already mentioned (p. 72, *supra*), is said to be a febrifuge.†

*Tamarindus indica*, L., the TAMARIND, a large tree, native of the tropical regions of the Old World, but cultivated largely in the West Indies, has a pulpy interior to its pods, which is mildly laxative. The Brown or Red Tamarinds of the West Indies come over preserved in syrup; the Black Tamarinds of the East Indies, without sugar or syrup.‡

*Copaifera Lansdorffi*, Desf., of Brazil, *C. officinalis*, L., of Central America, *C. Martii*, Hayne, of Guiana, and *C. guianensis*, Desf., are the sources of the oleo-resin known as BALSAM OF COPAIBA.§

*Hardwickia pinnata*, Roxb., a nearly allied large tree of Southern India, yields an oleo-resin used for Copaiba.‖

*Erythrophlœum guineense*, Don, is the 'Sassy Bark,' 'Mancona,' or 'Casca,' recently recommended in diseases of the heart and lungs. The tree is a native of Sierra Leone, and the bark is used in native ordeals.¶

The gums derived from the genus *Acacia*, and the astringent Cutch extracted from its duramen, are mentioned in the sequel.

* 'Pharmacographia,' pp. 195-7; 'Kew Museum Guide,' p. 49; Bentley and Trimen, ii, pl. 87.
† Christy, 'New Commercial Plants,' No. 8, p. 40.
‡ 'Pharmacographia,' pp. 197-200; Bentley and Trimen, ii, pl. 92.
§ Bentley and Trimen, ii, pl. 93; 'Kew Museum Guide,' p. 52.
‖ 'Pharmacographia,' p. 205.
¶ Lauder Brunton, 'Lancet,' December 2, 1876; June 26, 1880.

## ROSACEÆ.

The use of the oils extracted from Sweet and Bitter Almonds (*Amygdalus communis*, L., vars. *dulcis* and *amara*, DC.) has been already mentioned (p. 59, *supra*). Under the influence of the organic ferment ' emulsin,' or ' synaptase,' a form of vegetable casein, the crystalline substance, ' amygdalin,' present in Bitter Almonds, decomposes into prussic acid, oil of bitter almonds, and a glucose, as shown by Liebig and Wöhler in 1837 :*

$$C_{20}H_{27}NO_{11} + 2H_2O = HCN + C_7HO + C_{12}HO_{12}$$

Amygdalin    Water    Prussic    Oil of    Dextro-
                        Acid   Bitter Almonds   glucose.

*Prunus serotina*, Ehrhart, the American WILD BLACK CHERRY, the bark of which is there esteemed for its tonic and sedative properties, was introduced to notice by Professor Bentley, in 1863.†

*P. Lauro-cerasus*, L., the CHERRY-LAUREL, was admitted into our Pharmacopœia in 1839, for the manufacture of cherry-laurel water, now superseded by prussic acid.

*Hagenia abyssinica*, Willd., ' KOSO,' the flowers of which contain an acrid bitter resin, known as Kosin, has been used as a vermifuge. It was introduced here in 1850,‡ and admitted to the Pharmacopœia in 1864.

Reference will be made in the sequel to the OTTO or ATTAR OF ROSES, prepared from the petals of

* ' Pharmacographia,' pp. 220-1 ; Bentley and Trimen, ii, pl. 99.
† ' Pharm. Journ.,' v (1864), p. 97 ; Bentley and Trimen, ii, pl. 97.
‡ Pereira, ' Pharm. Journ.,' x (1851), p. 15 ; Bentley and Trimen, ' Medicinal Plants,' ii, pl. 102 ; ' Pharmacographia,' pp. 228-30.

*Rosa-gallica*, L., *R. centifolia*, L., and *R. damascena*, Mill., chiefly in Roumelia. Rose-petals are largely distilled for rose-water, for which purpose roses are grown at Mitcham and elsewhere, in Holland, and round Paris, and an infusion is commonly used as a vehicle for other medicines.

*Cydonia vulgaris*, Pers., the QUINCE, has a mucilaginous testa to its pips, which is said to be useful in dysentery, and as a demulcent in skin and eye complaints.

### HAMAMELIDEÆ.

*Hamamelis virginiana*, L., is said to be the source of the popular POND'S EXTRACT OF WYCH-HAZEL.

*Liquidambar orientalis*, Miller, a native of the South-west of Asia Minor, was shown, in 1841, to be the botanical origin of LIQUID STORAX, a soft viscid resin, now used mainly in perfumery.*

### MYRTACEÆ.

*Eucalyptus globulus*, Labil., and other species, are supposed to possess febrifugal properties in their leaves, which have been recommended as cigars in asthma. They yield an antiseptic oil, somewhat similar to Cajeput.†

*Syzygium Jambolanum*, DC., a large East India tree, has seeds which are recommended in diabetes.‡

---

* D. Hanbury, 'Pharm. Journ.,' xvi (1857), pp. 417, 461 ; and iv (1863), p. 436 ; Bentley and Trimen, ii, pl. 107.

† Bentley and Trimen, ii, pl. 109. C. R. Wattel, 'Pharm. Journ.,' iii (1872), pp. 22, 43 ; iv (1873), p. 494 ; vi (1876), p. 912 ; and vii (1876), p. 90.

‡ Christy, 'New Commercial Plants,' No. 10, p. 63.

*Melaleuca Leucadendron*, L., var. *minor*, Smith, a native tree of the Indian Archipelago, is the plant from the leaves of which the rubefacient and stimulant CAJEPUT-OIL is obtained by distillation. It is prepared mainly in Celebes, Bouro, and Amboyna, and consumed chiefly in India.*

CLOVES and MOTHER CLOVES, only important as spices, have been already mentioned (p. 66, *supra*). CLOVE-STALKS, another product of *Eugenia caryophyllata*, Thunb., are used to adulterate ground cloves. The essential oil is used in pills and for toothache.†

PIMENTO (*Pimenta officinalis*, Lindl.) yields a distilled water, frequently prescribed.‡ (See also p. 66, *supra*).

The cork both of the stem and of the root of the POMEGRANATE (*Punica Granatum*, L.) is officinal in India.§

### CUCURBITACEÆ.

*Ecballium Elaterium*, A. Richard (*Momordica*, L.), the SQUIRTING CUCUMBER, a native of Southern Europe, cultivated on a small scale at Mitcham and Hitchin, yields the powerful cathartic ELATERIUM, the juice round its seeds containing the crystallizable principle Elaterin, $C_{20}H_{28}O_5$.‖

*Citrullus Colocynthis*, Schrader (*Cucumis*, L.), COLOCYNTH, a plant more widely distributed though the drier regions of the Old World, contains Colocynthin, $C_{56}H_{84}O_{23}$, and is similarly employed.¶

---

* Bentley and Trimen, ii, pl. 108.      † *Ibid.*, ii, pl. 112.
‡ *Ibid.*, ii, pl. 111.                  § *Ibid.*, ii, pl. 113.
‖ *Ibid.*, ii, pl. 115.                  ¶ *Ibid.*, ii, pl. 114.

## TURNERACEÆ.

*Turnera diffusa,* Willd., and its variety, *aphrodisiaca,* DAMIANA, is recommended as a tonic for dyspepsia. It contains gum, tannin, resins, and a volatile oil.*

## PAPAYACEÆ.

*Carica Papaya,* L., the Papaw, a fruit believed to be of American origin, has most interesting properties, which, since about 1878, have been utilized in medicine. It contains an essential principle, ' Papaine,' analogous to the animal ferment trypsine, which gives it a rapid solvent action, that has suggested its successful use in dyspepsia, diphtheria, etc.†

## UMBELLIFERÆ.

*Hydrocotyle asiatica,* L. In the Pharmacopœia of India, and admitted to have some alterative tonic properties ; is suggested for use in scrofula.‡

*Œnanthe crocata,* L., is said to be a specific in epilepsy.§

*Conium maculatum,* L., the HEMLOCK, a British plant, contains in its leaves, and more especially in its fruits, the alkaloid Conia, $C_8H_{15}N$, a poison which is employed as a sedative and antispasmodic with other allied substances. The fruits were only introduced into our Pharmacopœia in 1864.||

* 'Pharm. Journ.,' vi (1875-6), pp. 24, 423, 581.
† *Ibid.,* ix (1878), p. 449 ; S. P. Oliver, *ibid.,* x (1879), p. 68. A. Wurtz and Bouchut, *ibid.,* p. 283 ; T. Peckolt, *ibid.,* pp. 343. 383.
‡ Christy, ' New Commercial Plants,' No. 10, p. 97 ; Bentley and Trimen, ii, pl. 117.
§ Bentley and Trimen, ii, pl. 124 ; Christy, *op. cit.,* p. 41.
|| Bentley and Trimen, ii, 118.

CARAWAYS, the half-fruits of *Carum Carui*, L., are mainly employed as a spice (see p. 66, *supra*); but the essential oil or distilled water is used in perfumery and as a stimulant in medicine.

*Carum copticum*, Benth., the AJOWAN of India (= *C. Ajowan*, Benth. and Hook. fil. = *Ptychotis Ajowan*, DC.) is now a chief source of the stimulant and carminative stearoptene THYMOL, $C_{10}H_{13}HO$, also obtained from *Thymus vulgaris*, L., and *Monarda punctata*, L. The fruits are mainly exported from Calcutta to Leipzig. Thymol was admitted into our Pharmacopœia in 1885.*

The fruits of FENNEL (*Fœniculum capillaceum*, Gilib.), which contain the stearoptene Anise-camphor or Anethol, $C_{10}H_{12}O$, are distilled, but chiefly for cattle medicines and cordials (see p. 64, *supra*). There are several varieties in commerce.† The fruits of ANISE (*Pimpinella Anisum*, L.) have a similar composition (see p. 65, *supra*) and are similarly employed.

*Ferula Sumbul*, Hook. fil., discovered in 1869 in the mountains south-east of Samarkand, is the source of SUMBUL, which was introduced into Russia about 1835 as a substitute for musk, was then used in cases of cholera, reached England about 1850, and was admitted into our Pharmacopœia as an antispasmodic in 1867.‡

Similarly *Ferula Narthex*, Boiss., *F. Scorodosma*, Benth., and allied species, producing the reputedly stimulant ASAFŒTIDA, have only become well known

---

* Spon, 'Encyclop. Indust. Arts,' p. 791 ; Bentley and Trimen, 'Medicinal Plants,' ii, pl. 120, 205, 208.

† 'Pharmacographia,' pp. 274-5 ; Bentley and Trimen, ii, pl. 123.

‡ Bentley and Trimen, ii, 129 ; 'Pharmacographia,' p. 278.

from the collections of Buhse, Bunge, Bellew, Aitchison, and other recent travellers in Persia and Afghanistan, whilst GUM SAGAPENUM is not yet certainly ascertained, but is believed to be from *F. persica*, Willd., or *F. Szowitsiana*, DC.*

GUM GALBANUM, an antispasmodic and stimulant expectorant, reaching Russia *viâ* Astrachan and Orenburg, but also obtained from Bombay and the Levant, is apparently the product of the Persian *F. galbaniflua*, Boissier and Buhse,† and *F. rubricaulis*, Boiss., and of *F. Schaïr*, Borszc, of Turkestan.

GUM AMMONIACUM, a similar substance, shipped from the Persian Gulf *viâ* Bombay, is the product of *Dorema Ammoniacum*, Don,‡ *D. Aucheri*, Boiss.,§ and, in Africa, of *Ferula tingitana*, L.||

OPOPANAX, a gum-resin largely used in perfumery, is attributed to *Opopanax Chironium*, Koch, a native of the Mediterranean area.

The distilled water of the fruits of DILL (*Anethum graveolens*, L.) and CORIANDER (*Coriandrum sativum*, L.),¶ already referred to (pp. 65 and 66, *supra*), are carminative, and those of CUMIN (*Cuminum Cyminum*, L.)** are used to a considerable extent in veterinary practice.

### RUBIACEÆ.

*Uncaria Gambier*, Roxburgh, the astringent extract

---

* 'Kew Museum Guide,' p. 77.   E. M. Holmes, 'Pharm. Journ.,' 1888.
† Bentley and Trimen, ii., pl. 128.   ‡ *Ibid.*, pl. 131.
§ *Ibid.*, pl. 130.   || 'Kew Museum Guide,' p. 75.
¶ Bentley and Trimen, ii., pl. 134.   ** *Ibid.*, pl. 134.

from the shoots of which is known as 'GAMBIER,' is mainly important in dyeing and tanning.*

*Sarcocephalus esculentus*, Afzelius, 'DOUNDAKE,' or (?) AFRICAN PEACH-ROOT, though often mixed with bark of species of *Morinda* and *Cochlospermum*, yields extracts more powerfully febrifuge and sedative than Berberine, and giving an old-gold colour. It is recommended as a tonic and antiperiodic, and possibly as a quinine substitute.†

CINCHONA. Nothing in the modern history of Materia Medica equals in importance the identification of the various species of *Cinchona* that yield QUININE and other valuable alkaloids, and the introduction of these plants into other countries. There is no record of this invaluable drug before the year 1630. In 1638 the Countess Ana of Chinchon, wife of the viceroy of Peru, was cured of a fever by it, and it became known in Spain as 'Polvo de la Condesa,' the Countess' Powder. Jesuits introduced it to Brussels about 1650 as Palo de Calenturas, Lignum febrium, China febris, or Pulvis febrifugus, and a few years later it became known in England as Jesuits' Powder, being admitted into the Pharmacopœia in 1677 as Cortex Peruanus, Peruvian Bark. The reckless destruction of the trees for the sake of the bark having resulted in its extermination in many of its native habitats, directed attention to its cultivation in other countries. In 1872, 28,450 cwt. of Peruvian bark was imported. Hugh Algernon Weddell, in 1848, brought home

---

* Bentley and Trimen, ii., pl. 139.
† Christy, ' New Commercial Plants,' No. 8, p. 45, and No. 9, p. 7, with 2 plates ; ' Pharm. Journ.,' xvi, 1885, p. 49.

seeds, and in 1852 the Dutch began cultivation in Java. Royle recommended planting in the Nilghiris in 1839, and again in 1858, when he pointed out that the East India Company were spending £7,000 a year for bark and £25,000 for quinine. Mr. Clements Markham went to Peru and secured the assistance of the botanist Richard Spruce, of John Weir, and of Robert Cross in 1860. From that year plants have been raised at Kew, and from the year following at Ootacamund, then under the charge of W. G. McIvor. They have since been largely planted at Darjeeling, in the Kangra Valley, in Ceylon, Burma, Mauritius, Jamaica, and Trinidad. Indian Cinchona bark first came into the English market in 1867. Ceylon now exports a very large amount. This dispersal of these valuable plants, though largely carried out by means of seeds, was much facilitated by Mr. N. B. Ward's invention of the cases which are always known by his name.

Though there are numerous other species, varieties and hybrids, those chiefly grown are the original *C. officinalis*, L., yielding PALE CINCHONA, CROWN or LOXA BARK ; *C. Calisaya*, Weddell, discovered by that botanist in 1847, and its variety, *Ledgeriana*, Moens, yielding YELLOW CINCHONA, or CALISAYA BARK, and *C. succirubra*, Pavon, yielding RED CINCHONA BARK. This last is largely cultivated as a source of cinchonidine, and from it the cheap febrifuge made in India is prepared.* All of these are trees

---

* 'Pharmacographia,' pp. 302-31, where a bibliography is given ; 'Flückiger, F. A., and Power, F. B., 'The Cinchona Barks,' 1884. Bentley and Trimen, ii., pl. 140-143.

from 20 to 40 feet high, growing wild on the Andes at altitudes of from 2,500 to 7,500 feet.

*Remijia Purdieana*, Willd., and *R. pedunculata*, Trian., from Colombia, have of late years been imported in enormous quantities as sources of quinine etc.,* under the name of CUPREA BARK.

IPECACUANHA, the root of a small Brazilian shrub, *Cephaëlis Ipecacuanha*, A. Richard, a useful emétic, and the only known specific for dysentery, was introduced into Europe in 1672. We import about 65,000 lb., valued at nearly £15,000, annually. The plant has been introduced into Northern India from the Edinburgh Botanic Garden since 1866.† Mention may here be made of the Violaceous *Ionidium Ipecacuanha*, Vent., which, with allied species constituting the 'Poaya blanca' of Brazil, has been sold as 'WHITE IPECACUANHA.'

### VALERIANACEÆ.

*Valeriana officinalis*, L., a British species, the rhizome of which yields a volatile oil, possessing antispasmodic properties, is cultivated near Chesterfield, Derbyshire ; in Holland ; and in the north-eastern United States.‡

### COMPOSITÆ.

*Inula Helenium*, L., ELECAMPANE, a rare British perennial, has an aromatic tonic root, now mainly used in Absinthe (see p. 63, *supra*), and in veterinary medicine. It gives its name to Inulin, the soluble,

* Christy, 'New Commercial Plants,' No. 8, p. 41 ; and 'Pharm. Journ.,' 1884.
† 'Pharmacographia,' pp. 331-7. Bentley and Trimen, ii., pl. 145.
‡ *Ibid.*, pl. 146.

crystallizable isomer of starch, characteristic of the Compositæ.*

*Anacyclus Pyrethrum*, DC., PELLITORY, a native of Algeria, which is perennial, and *A. officinarum*, Hayne, GERMAN PELLITORY, an annual, cultivated in Germany, have roots which are employed as a rubefacient and as a gargle.†

*Anthemis nobilis*, L., CHAMOMILE, a perennial British plant, is cultivated at Mitcham, the inflorescence of the cultivated form having nearly all its florets white 'ray' flowers, *i.e.*, being 'double.' The inflorescence is infused as a bitter stomachic and tonic tea, or the whole plant is distilled for a carminative oil.‡

*Artemisia pauciflora*, Weber, SANTONICA, a native of the Kirghiz steppes, yields the valuable anthelmintic known as WORMSEED, which consists of the unopened flower-heads or capitula, and contains a crystalline body known as Santonin, $C_{15}H_{18}O_3$, discovered in 1830.§ The wood of other species of this genus was formerly used for the same purpose, whence the name WORMWOOD, and in fever before the introduction of Cinchona. The common British MUGWORT (*A. vulgaris*, L.) is a rustic remedy for rheumatism.

*Arnica montana*, L., a native of Central Europe, yields a tincture used externally for bruises, sprains, etc.‖

*Taraxacum officinale*, Wiggers, the DANDELION, a widely distributed weed, yields a milky latex,

---

* Bentley and Trimen, iii., pl. 150.  † *Ibid.*, pl. 151, 152.
‡ *Ibid.*, pl. 154.  § *Ibid.*, pl. 157.  ‖ *Ibid.*, pl. 158.

chiefly from the root, which is a mild laxative and tonic, and has been in use for centuries.

*Mikania Guaco,* H. B., GUACO, a cure for snake-bite, popular in South America and the West Indies, has been recommended as a febrifuge and anthelmintic.*

*Siegesbeckia orientalis,* L., 'Herbe de Flacq,' 'Guérit vite,' is suggested for use as an alterative, etc. It is used in Mauritius, and contains a principle known as Darutyne.†

*Parthenium Hysterophorus,* L., of the West Indies, yields Parthenine, a suggested quinine substitute.‡

*Lactuca altissima,* Bieb., possibly only a variety of *L. Scariola,* L., a native of the Caucasus, now cultivated in Auvergne, the species just named, *L. virosa,* L., and *L. sativa,* L., the Garden Lettuce, are the sources of LACTUCARIUM, a sedative produced by the Fairgrieve family near Edinburgh from about 1844, near Clermont-Ferrand, Auvergne, from 1841, and near Zell, in Rhenish Prussia, from 1845.§

### LOBELIACEÆ.

*Lobelia inflata,* L., INDIAN TOBACCO, a North-American species, in small doses is an expectorant, useful in spasmodic asthma; in larger doses a powerful emetic, or even an acro-narcotic poison.‖

* P. L. Simmonds, 'Pharm. Journ.,' x (1851), p. 534 ; xiii (1854), p. 412 ; J. G. Baker, *ibid.,* xi (1880), pp. 471, 369.
† Christy, 'New Commercial Plants,' No. 9, p. 49, and No. 10, p. 85.
‡ *Ibid.,* No. 10, p. 99.
§ 'Pharmacographia,' p. 345 ; 'Pharm. Journ.,' viii (1877), p. 202. Bentley and Trimen, iii., pl. 160, 161.
‖ *Ibid.,* pl. 162.

### ERICACEÆ.

*Arctostaphylos Uva-ursi,* Sprengel, the BEARBERRY, a northern mountain evergreen, has astringent leaves, useful as a tonic in affections of the bladder.*

### EBENACEÆ.

*Diospyros Embryopteris,* Pers., the GAUB of India, has an astringent pulp in the unripe fruit, used in India in dysentery.†

### STYRACEÆ.

*Styrax Benzoin,* Dryander, is believed to be the source of GUM BENZOIN or GUM BENJAMIN, of Sumatra and Siam, which is used in bronchitis, etc., and forms a principal ingredient in 'FRIAR'S BALSAM,' though it is mainly consumed as incense in the Greek Church.‡

### OLEACEÆ.

*Fraxinus Ornus,* L., the FLOWERING or MANNA ASH of Sicily and Southern Italy, is the chief source of the mild laxative, MANNA, which consists mainly of a sugar known as Mannite, $C_6H_{14}O_6$.§

OLIVE-OIL, the produce of *Olea europæa,* L., though employed in medicine, will be more fully alluded to in the sequel.

### APOCYNACEÆ.

*Carissa xilopicron,* Thouars, in an alcoholic extract of the root,‖ *Geissospermum Vellosii,* Allem, which contains

* Bentley and Trimen, iii., pl. 163.　† *Ibid.,* pl. 168.
‡ 'Kew Museum Guide,' p.94. Bentley and Trimen, iii., pl. 169.
§ *Ibid.,* pl. 170.
‖ Christy, ' New Commercial Plants,' No. 10, p. 103.

Pereirine, *Alstonia scholaris*, R. Br., 'DITA BARK,' the bark of a tree well known in the tropics of the Old World,* *A. constricta*, F. von Muell., its Australian congener, now much used in the United States as 'QUEENSLAND FEVER-BARK,'† and *Aspidosperma Quebracho*, the 'WHITE QUEBRACHO BARK' of Chili, which contains Aspidospermine, are all members of this order, which have been recommended as tonic antiperiodic febrifuges instead of quinine.

*Strophanthus hispidus*, DC., or *S. Kombe*, Oliver, the 'WANIKA ARROW POISON' of East Africa, is stated‡ to be almost a specific in fatty degeneration and other cardiac affections, the root being the part employed.

*Holarrhena antidysenterica*, Wall., is the 'MANOIS REMEDY' for chronic dysentery in Mauritius. It yields a principle known as Conessine.§

*Gelsemium nitidum*, Michaux, 'CAROLINA JESSAMINE-ROOT,' has been admitted to the Pharmacopœia (1885), its rhizome being used in neuralgia, rheumatism, and fever, though over-doses seem dangerous.‖

ASCLEPIADEÆ.

*Hemidesmus indicus*, R. Br., INDIAN SARSAPARILLA, a twining shrub, is used mainly in Southern India.¶

* 'Pharm. Journ.,' xii (1853), p. 422 ; vi (1875), p. 142 ; 'Pharmacographia,' p. 378.   Bentley and Trimen, iii., pl. 173.
† 'Chemical News,' November 15th, 1878.
‡ Christy, 'New Commercial Plants,' Nos. 9 and 10, p. 7, with engravings of several species.
§ *Ibid.*, No. 10, p. 40, and ' Pharm. Journ.,' xvii (1886), p. 83.
‖ 'Pharm. Journ.,' iv (1874), p. 998 ; vi (1875-6), pp. 481, 521, 561, 601 ; Bentley and Trimen, iii, p. 181.
¶ 'Kew Museum Guide,' p. 97.   Bentley and Trimen, iii., pl. 174.

*Calotropis gigantea,* R. Br., and *C. procera,* R. Br., the MUDAR, Indian fibre-yielding shrubs, of which the latter is naturalized in North-east Africa and the West Indies, have an alterative tonic bark to their roots.*

*Tylophora asthmatica,* Wight and Arnott, COUNTRY IPECACUANHA, common in India and naturalized in Mauritius, is used as a substitute for Ipecacuanha.†

## LOGANIACEÆ.

*Spigelia anthelmintica,* the DEMERARA PINK-ROOT, is recommended as a vermifuge, a use to which its congener, *S. marilandica,* L., has been put for a century.‡

*Strychnos toxifera,* Schomb., together with *S. cogens,* Benth., *S. Schomburghii,* Klot., and other species in other districts, are the sources of the celebrated Wourali, or Curare poison of Guiana, which has been tried for rabies, and which, from its anæsthetic action, has proved of such value in physiological research.§

*Strychnos Nux-vomica,* L., indigenous to India, and *S. Ignatii,* Bergius, ST. IGNATIUS' BEAN, a native of the Philippines, contain in their seeds the bitter, highly poisonous, but tonic alkaloids STRYCHNINE, $C_{21}H_{22}N_2O_2$, and BRUCINE, $C_{23}H_{26}N_2O_4$.‖

---

\* Bentley and Trimen, iii., pl. 176. 'Catalogue of Indian Exhibits : Colonial and Indian Exhibition,' p. 103.

† Bentley and Trimen, iii., pl. 177.

‡ *Ibid.,* pl. 180. G. R. Bonyan, 'Pharm. Journ.,' v (1846), pp. 354-5.

§ T. Redwood in 'Pharm. Journ.,' iii (1843), p. 74 ; J. Moss, *ibid.,* viii (1877), p. 421 ; and Gustave Planchon, *ibid.,* xi (1880-1), pp. 469, 491, 754.

‖ Bentley and Trimen, iii., pl. 178, 179.

### GENTIANEÆ.

*Gentiana lutea*, L., yields the bitter tonic GENTIAN-ROOT, which has long been much used in medicine. It is collected in the mountains of Switzerland, and other parts of Central and Southern Europe.*

*Ophelia Chirata*, Griseb., 'CHIRETTA' of India, a valuable bitter tonic, was admitted to the Edinburgh Pharmacopœia in 1839. When cheap it is substituted for gentian in cattle foods.†

### CORDIACEÆ.

*Cordia Boissieri*, DC., ANACAHUITE-WOOD, was imported about 1860, from Tampico, in Mexico, as a remedy for consumption.‡

### CONVOLVULACEÆ.

*Convolvulus Scammonia*, L., SCAMMONY, a native of the Levant, has a milky juice in its long tap-root, which yields a gum-resin called JALAPIN, $C_{31}H_{50}O_{16}$, and has accordingly long been valued as a purgative. The same substance probably occurs in the KALA-DANA RESIN, obtained from the seed of *Pharbitis Nil*, Choisy, which is officinal in India, as it certainly does in the roots of *Ipomœa*.

The true JALAP is the tuber of *Ipomœa Purga*, Hayne, a native of the Mexican Andes, growing at altitudes of from 5,000 to 8,000 feet, near Xalapa, from which place it takes its name, and imported from Vera Cruz to the extent of over eighty tons annually.

* Bentley and Trimen, iii., pl. 182.
† *Ibid.*, pl. 183. R. Bentley, 'Pharm. Journ.,' v (1874), p. 481 ; 'Pharmacographia,' p. 392.
‡ 'Pharm. Journ.,' ii (1861), p. 407 ; iii (1861), p. 164 ; iv (1862), p. 271.

META

*Ipomœa simulans,* Hanbury, TAMPICO JALAP, is often, *I. orizabensis,* Ledanois, ORIZABA-ROOT, more rarely, imported in lieu of the true Jalap.*

#### SOLANACEÆ.

*Solanum paniculatum,* L., JURUBEBA, of Brazil, is suggested as a tonic and diuretic, in syphilis, etc.†

*Solanum Dulcamara,* L., the BITTERSWEET, or WOODY NIGHTSHADE of our hedge-rows, though used in the form of a decoction of the leaves and stalks, which have a taste at first bitter and then sweet, has no certain value.‡

CAPSICUMS, the fruits of *Capsicum fastigiatum,* Blume, and *C. annum,* L., already alluded to (p. 64, *supra*), are of use as digestives, and in gargles.§

*Atropa Belladonna,* L., the DEADLY NIGHTSHADE, not uncommon on our limestone hills, is a powerful narcotic poison. Its leaves, and especially its roots, contain the alkaloid Atropine, $C_{17}H_{23}NO_3$. This produces contraction of the iris, *i.e.,* dilation of the pupil of the eye, whence its use in ophthalmic medicine, and by the ladies of Venice in the sixteenth century, from which it derived the name ' Bella donna.'||

*Datura Stramonium,* L., the THORN-APPLE, apparently a native of the Caspian region,¶ *D. Tatula,* L., of America, *D. fastuosa,* L., and *D. alba,* Nees, both Indian plants, contain in their leaves and seeds

---

* 'Pharmacographia,' pp. 401-2. Bentley and Trimen, iii., pl. 185-187.

† Christy, 'New Commercial Plants,' No. 10, p. 101.

‡ Bentley and Trimen. iii., pl. 190.    § *Ibid.,* pl. 188, 189.

|| *Ibid.,* pl. 193.    ¶ *Ibid.,* pl. 192. 'Pharmacographia,' pp. 412-16.

the narcotic poison Daturine, an isomer of Atropine, valuable as a sedative, and therefore the leaves are employed as cigars in asthma.*

*Hyoscyamus niger*, L., HENBANE, a British plant, cultivated on a small scale for medicinal purposes, is sedative, anodyne, or hypnotic, and is used instead of opium. It contains an alkaloid, Hyoscyamine, $C_{15}H_{23}NO_3$, which acts on the iris like Atropine.†

TOBACCO, the dried leaves of *Nicotiana Tabacum*, L., and *N. rustica*, L., natives of America, is but little used medicinally, though its active principle, the volatile oily alkaloid Nicotine, $C_{10}H_{14}N_2$, is a powerful narcotic poison. It is produced in the drying leaf by a fermentative process, but is mainly destroyed when the tobacco is burnt. The enormous growth of the use of this substance in Great Britain, as in the rest of the Old World, since its introduction in the sixteenth century, may be gauged from the facts that over £9,000,000 was paid on it as Customs duty in the year 1886-7, the tariff on unmanufactured tobacco being 3s. 6d. per lb., and that about 51,000,000 lb., or 1⅔ lb. per head of the population, was retained for home consumption during the same year. By far the greater part of this supply comes from the United States. SNUFF is prepared from the stalks and veins of the leaves. Tobacco has recently been experimentally cultivated in this country, without much success.‡

* De Candolle, ' Géographie Botanique,' ii (1855), p. 731.
† Bentley and Trimen, iii., pl. 194.
‡ Lock, C. G. W., ' Tobacco : Growing, Curing, and Manufacturing ; a Handbook for Planters in all Parts of the World,' 8vo., 1886.

*Duboisia Hopwoodi,* F. von Muell. ( = *D. myoporoides,* R. Br.), the broken leaves and twigs of which form PITURI, which is chewed by the Australian natives as a stimulating tonic, or intoxicant. It contains Hyoscyamine and Duboisine ; and, on its introduction to medicine by Dr. Bancroft, about 1873, proved to be a powerful mydriastic, one part in 32,000 sufficing to dilate the pupil.*

*Fabiana imbricata,* Ruiz and Pavon, of Chili and the Argentine, contains FABIANINE, and has proved serviceable in cystitis.†

### SCROPHULARIACEÆ.

*Digitalis purpurea,* L., the FOXGLOVE, a favourite British flowering plant, was mainly introduced into medicine by Withering.‡ It is a powerful poison, reducing the action of the heart ; but its dried leaves are used as a sedative and diuretic.

### ACANTHACEÆ.

*Andrographis paniculata,* Nees, the KARIYÁT of India, is used, like Chiretta, Gentian, or Quassia, as a stomachic bitter tonic.§

*Hygrophila spinosa,* Vel., is recommended as a diuretic.‖

### SESAMEÆ.

*Sesamum indicum,* DC., indigenous to India, but

---

* G. Bennett, ' Pharm. Journ.,' iv (1873), p. 184 ; vii (1877), p. 878 ; viii (1878), pp. 705, 720 ; ix (1879), p. 638.
† Christy, No. 10, p. 108.
‡ Withering, William, 'Account of the Foxglove,' 8vo., Birmingham, 1785. Bentley and Trimen, iii., pl. 195.
§ *Ibid.,* pl. 197.    ‖ Christy, No. 10, p. 111.

now cultivated in Southern Europe, has oily seeds from which is expressed GINGELY or SESAMÉ-OIL, the chief oil of India, a useful substitute for olive-oil, now largely used in soap-making, especially at Marseilles. Further reference will be made to it in Part III.

## LABIATÆ.

An order of aromatic herbs containing volatile oils holding in solution hydrocarbons analogous to Camphor, known as stearoptenes. Several of them have long been cultivated at Mitcham, Surrey; Hitchin, Herts; and Market Deeping, Lincolnshire.

*Mentha arvensis*, DC., vars. *piperascens* and *glabrata*, and *M. piperita*, Sm., are the sources of MENTHOL, or CHINESE OIL OF PEPPERMINT, $C_{10}H_{18} + H_2O$, which first came into commerce about 1872, and proved useful in neuralgia. It was admitted to the Pharmacopœia in 1885.*

The cultivation of PEPPERMINT (*M. Piperita*, Hudson), for its essential oil, has already (p. 67, *supra*) been mentioned. The common MINT or SPEARMINT (*M. viridis*, L.) of our kitchen-gardens is similarly cultivated and distilled, and the oil is employed in the United States as a perfume by soap-makers.†

*M. Pulegium*, L., PENNYROYAL, another British species, is distilled as a carminative and anti-spasmodic.‡

*Lavandula vera*, DC., LAVENDER, a native of South-west Europe, extensively cultivated at the above-

---

* Bentley and Trimen, iii., pl. 203.    † *Ibid.*, pl. 202
‡ *Ibid.*, pl. 201.

mentioned English localities, yields, mainly from the flowers, the favourite perfume, Oil of Lavender, that prepared at Mitcham being by far the best. The allied species *L. Spica*, DC., and *L. Stœchas*, L., yield the less fragrant OIL OF SPIKE, which is used in china-painting.*

*Thymus vulgaris*, L., the GARDEN THYME of our kitchen-gardens, a native of Southern Europe, is distilled in the South of France for the sake of its essential oil, an external stimulant containing the camphor Thymol, $C_{10}H_{14}O$.† This substance is now prepared mainly from Ajowan fruits, *Ammi Copticum*, L. (see p. 96, *supra*).

*Rosmarinus officinalis*, L., ROSEMARY, a native of the Mediterranean area, has long been in cultivation with us, for the sake of its flowers and the volatile oil distilled from them. This is used as a perfume and as an external stimulant, being supposed to promote the growth of the hair. It is largely produced in the island of Lesina, Dalmatia, and exported from Trieste.‡

*Orthosiphon stamineus*, Benth., of Java, is recommended in affections of the kidneys and bladder.§

*Marrubium vulgare*, L., the HOREHOUND, is now grown in California, to flavour honey for colds.||

### PLANTAGINEÆ.

The mucilaginous seeds of *Plantago decumbens*, Forsk., ISPAGHÚL, or SPOGEL, are used in India as a demulcent drink, especially in dysentery.¶

* Bentley and Trimen, iii., pl. 199.    † *Ibid.*, pl. 205.
‡ *Ibid.*, pl. 207.                     § Christy, No. 10, p. 104.
|| Bentley and Trimen, iii., pl. 210.   ¶ *Ibid.*, pl. 211.

### POLYGONACEÆ.

*Rheum palmatum*, L., the real source of the long-used stomachic and purgative RUSSIAN or TURKEY RHUBARB, was first found wild by Colonel Prejevalsky, in the extreme North-west of China, in 1872-3.*

*R. officinale*, Baillon,† found in South-east Thibet by French missionaries, has been grown at Bodicott, near Banbury, since 1873, where also *R. Rhaponticum*, L., the species used for culinary purposes, is also grown for its roots (see p. 56, *supra*). Our imports of medicinal Rhubarb amount to about 350,000 lb. annually.‡

### MONIMIACEÆ.

*Atherosperma moschata*, Lab., of Australasia, the AUSTRALIAN SASSAFRAS bark, is used in asthma and bronchitis as a sedative,§ and the leaves of *Peumus Boldus*, Molin., the BOLDO, of Chili, have been recently imported for use in liver complaints, syphilis, etc.‖

### MYRISTICEÆ.

NUTMEGS, the seeds of *Myristica fragrans*, Houttuyn, which grows chiefly on the Banda Islands, Jilolo, Celebes, and Amboyna, though used chiefly as

* Bentley and Trimen, iii., pl. 214.
† *Ibid.*, pl. 213. 'Pharm. Journ.,' iii (1872), p. 301, and iv (1874), p. 690. See also vol. viii (1878), p. 588.
‡ 'Kew Museum Guide,' p. 107 ; 'Pharmacographia,' pp. 442-51.
§ Christy, No. 4, p. 46, and No. 10, p. 38.
‖ Bentley and Trimen, iii., pl. 217.

a condiment (see p. 66, *supra*), are employed in medicine as a stimulant. The seeds are imported *viâ* Batavia and Singapore. They are noticeable structurally for the 'ruminated' or marbled appearance seen in section, due to the folding of the inner seed-coat into the albumen, and for the 'aril,' or Mace.*

## LAURINEÆ.

*Sassafras officinale*, Nees, *Laurus Sassafras*, L., the SASSAFRAS of the United States, is aromatic, especially in the wood, root, and root-bark, and is distilled chiefly in its native country to flavour soap, tobacco, and drinks.† It is is said to be sudorific.

*Cinnamomum Camphora*, Fr. Nees and Ebermaier, *Laurus Camphora*, L., the CAMPHOR LAUREL, yields the true CAMPHOR, $C_{10}H_{16}O$, a volatile, crystalline sublimate from the wood, prepared in Japan and Formosa, long valued as a perfume and medicine. We import over 100 tons of refined, and 600 tons of unrefined, camphor annually. Camphor is a stimulant, and is considerably used in an alcoholic solution.

*C. zeylanicum*, Breyne, a variable species, is the main source of CINNAMON ; and other species, such as *C. Cassia*, Bl., and *C. iners*, Reinw., of CASSIA LIGNEA, both which are used as stimulants and cordials, but mainly as spices (see p. 67, *supra*).‡

*Cinnamomum Cassia*, Bl., of Southern China, is generally stated to be the source of CASSIA LIGNEA, or CASSIA BARK, which, though used as a cordial and

* Bentley and Trimen, iii., pl. 218.
† *Ibid.*, pl. 220. 'American Journal of Pharmacy,' 1871, p. 470.
‡ Bentley and Trimen, iii, pl. 223, 224.

stimulant in medicine, is more largely consumed as a spice; but Sir J. D. Hooker, in his 'Himalayan Journals,'[*] describes several other species as the sources of that produced in the Khasya mountains.

*Nectandra Rodiæi*, Schomburgk, the well-known GREENHEART, or BEBEERU of British Guiana, contains in its bark an alkaloid Bebeerine (now shown to be identical with Buxine, $C_{18}H_{21}NO_3$), discovered by Hugh Rodie, a naval surgeon, in 1835. It was examined by Dr. Maclagan in 1843, and has been admitted to the Pharmacopœia as a febrifuge. It forms the main ingredient of 'WARBURG'S DROPS,' a popular medicine in India.[†]

## THYMELÆACEÆ.

*Daphne Mezereum*, L., MEZEREON, a rare British species, contains in its acrid bark a glucoside, Daphnin, $C_{31}H_{34}O_{19}$, and has been prepared as an ethereal extract for a stimulating liniment.[‡] The bark itself is used as a vesicant.

## URTICACEÆ.

*Cannabis sativa*, L., the HEMP, a native of Asia, yields various products of a resinous character, chiefly valued in Oriental medicine. 'GUAZA,' or 'GUNJAH,' occasionally comes into the English drug market, and consists of the dried flowering shoots of the plant.

BHANG, or HASHÍSH, the dried resinous leaves and

* Vol. ii, p. 303, ed. 1855.
† Maclagan, 'Pharm. Journ.,' iii (1843), p. 177; H. Rodie, *ibid.*, iv (1844), p. 281; Bentley and Trimen, iii, pl. 219.
‡ 'Pharmacographia,' p. 487. Bentley and Trimen, iii, pl. 225.

fruits, is smoked in India, as is CHARAS, the crude resin. All these products are narcotic.*

*Humulus lupulus*, L., the HOP, a British-species, yields the bitter principle Lupulin, $C_{32}H_{50}O_7$, first isolated in 1863, together with wax and resin (see p. 61, *supra*). Hops warmed in pillows form a useful sedative, and an extract of the bitter principle is now largely used as a tonic drink.†

### ULMACEÆ.

*Ulmus fulva*, Michaux, the SLIPPERY ELM of North America, contains in its inner bark a mucilage which causes it to be powdered for poultices.‡

### EUPHORBIACEÆ.

*Euphorbia resinifera*, Berg, was described in 1863 as the source of the once used GUM EUPHORBIUM, which is still employed as a vesicant and in veterinary practice. It is a native of the lower slopes of the Atlas Mountains.§

*E. pilulifera*, L., the AUSTRALIAN ASTHMA HERB, has been lately recommended in bronchial cases.‖

*E. Drummondii*, Boiss, yields the anæsthetic DRUMINE, an alkaloid similar to Cocaine.¶

*Croton Eluteria*, Bennett, in 1859 was shown** from

---

* *Ibid.*, p. 493; *Ibid.*, iv, pl. 231. Cooke, 'Seven Sisters of Sleep.'
† Bentley and Trimen, iv, pl. 230.
‡ *Ibid.*, iv, pl. 233.
§ *Ibid.*, iv, pl. 240.  'Pharmacographia,' p. 502.
‖ 'Pharm. Journ.,' xvii (1886), p. 144.
¶ Christy, 'New Commercial Plants,' No. 10, p. 95.
** 'Proc. Linn. Soc., Bot.,' iv. (1860), 29, and Daniell, 'Pharm Journ.,' iv (1863), pp. 144, 226.  Bentley and Trimen, iv, pl. 238.

specimens collected in the Bahamas, by Daniell, in 1857-8, to be the true source of CASCARILLA BARK, which is imported to a considerable extent as a tonic. A spurious bark sometimes sent over is suggested by Mr. E. M. Holmes to be that of *C. lucidus*, L.*

*C. niveus*, Jacquin (*C. Pseudo-China*, Schlecht.), the 'Quina blanca,' of Mexico, yields COPALCHI BARK, occasionally imported as a quinine substitute.†

*C. Tiglium*, L., yields from its seeds the powerfully cathartic and rubefacient CROTON OIL. The tree is a native of the Malabar Coast, and the seeds are mainly imported from Bombay.‡

*Ricinus communis*, L., the CASTOR OIL, a native of India, long cultivated in Europe, but only admitted to the London Pharmacopœia in 1788, contains oil to the extent of about half the weight of its kernels. We import over 1,800 tons of this oil annually, about two-thirds from India and the rest chiefly from Italy. It is a valuable mild purgative; but is also used in soap-making.§

*Mallotus philippinensis*, Müll. Arg. (=*Rottlera tinctoria*, Roxb.), a tree common throughout the Madras Presidency, has its capsules covered with ruby-like glands which constitute the powder known as KAMALA, or WARS. It is used as a dye for silk, and during the last thirty years as a tænifuge. It was admitted to our Pharmacopœia in 1864.∥

* 'Pharm. Journ.,' v (1874), p. 810.
† *Ibid.*, ix (1850), p. 463; and xiv (1855), p. 319.
‡ Bentley and Trimen, iv, pl. 239.
§ *Ibid.*, iv, pl. 237.
∥ Smith, 'Domestic Botany,' p. 263; 'Pharmacographia,' p. 515. Bentley and Trimen, iv, pl. 236.

## PIPERACEÆ.

*Piper angustifolium*, Ruiz and Pavon (=*Artanthe elongata*, Miquel), MATICO, of northern South America, was introduced to notice as a styptic in Europe in 1839.* In 1863, Professor Bentley recognised that *P. aduncum*, L., was equally efficacious.†

*P. Cubeba*, L. fil., *Cubeba officinalis*, Miquel, CUBEBS, a native of Java, Borneo, and Sumatra, once used as a spice, has, since 1815, been successfully employed in gonorrhœa.‡ The small, dried, unripe fruits contain a resin, which is the active part of the drug. We import nearly 60 tons of Cubebs annually, *viâ* Singapore.

*P. methysticum*, Forst., the KAVA ROOT, has been recommended by Prof. Gubler as a sudorific, and as a substitute for Cubebs.§

## ARTOCARPACEÆ.

*Dorstenia Contrajerva*, L., CONTRAYERVA ROOT, of Brazil, is a rhizome imported to this country, and having a reputation as a cure for snake-bite.‖

* Jeffreys, 'Remarks on the Efficacy of Matico,' London, 1845.
† 'Pharm. Journ.,' v (1864), p. 290. See also Von Martius, *ibid.*, ii (1843), p. 660; *ibid.*, v (1874), p. 523; J. Marcotte, 'Etude générale de Matico,' Paris, 1863; and Bentley and Trimen, iv, pl. 242.
‡ Crawfurd, 'Edinburgh Medical and Surgical Journal,' xiv (1818), p. 32; 'Pharmacographia,' pp. 526-30. Bentley and Trimen, iv, pl. 243.
§ 'Pharm.' Journ.,' August, 1876, p. 149; Christy, No. 10, p. 93.
‖ Smith, 'Domestic Botany,' p. 230.

### ARISTOLOCHIACEÆ.

The rhizome of *Aristolochia serpentaria*, L., the VIRGINIAN SNAKE-ROOT, is now used, generally with Cinchona, as a stimulant tonic. *A. reticulata*, Nutt., TEXAN SNAKE-ROOT, is also used, as are also its congeners in Peru, Brazil and India.*

### CUPULIFERÆ.

The bark of our COMMON OAK (*Quercus Robur*, L.) and the ALEPPO or TURKEY GALLS, produced by the puncture of a gnat on the shoots of *Q. infectoria*, Oliv., are important sources of the astringent Tannic and Gallic Acids.†

### SALICACEÆ.

One of the most important discoveries of late years has been that of SALICYLIC ACID ($HC_7H_5O_3$), as a specific in many rheumatic affections. It is admitted to the Pharmacopœia as obtained from *Betula lenta*, L., or *Gaultheria procumbens*, L. ; but the active principle, SALICINE, $C_{13}H_{18}O_7$, now much used as a quinine substitute, is obtained from the barks of *Salix Russelliana*, Sm., *S. alba*, L., *S. Caprea*, L., *S. fragilis*, L., *S. pentandra*, L., *S. purpurea*, L., and other species of that genus and of *Populus*.

### SANTALACEÆ.

*Santalum album*, L., the SANDAL-WOOD of India ; *S. Freycinetianum*, Gaudin, and *S. pyrularium*, A. Gray, of the Sandwich Islands ; *S. Yasi*, Seemann, of Fiji ; *S. Austro-caledonicum*, Vieill., of New Caledonia ;

* Bentley and Trimen, iv, pl. 246.
† *Ibid.*, iv, pl. 248, 249.

and *Fusanus acuminatus*, R. Br., of Australia, though mainly used for cabinet-work, incense, and perfumery, yield an essential oil, lately used as a substitute for Copaiba.*

ZINGIBERACEÆ.

*Zingiber officinale*, Roscoe, GINGER, a native of Asia, now largely cultivated in Jamaica, though mainly used as a condiment (see p. 65, *supra*), is a useful stomachic. We import some 1,600 tons of its rhizomes annually.†

*Elettaria Cardamomum*, Maton, the true or MALA-BAR CARDAMOM, a native of Southern India, has long been valued in India and in Northern Europe, its aromatic seeds, which contain a camphor isomeric with that of turpentine, $C_{10}H_{16}(H_2O)^3$, being used mainly as a condiment (see p. 67, *supra*). The seeds of various species of *Amomum* are substituted for it in Asia.‡

*Amomum Melegueta*, Roscoe, GRAINS OF PARA-DISE, the production of which gives its name to the Grain Coast of West Tropical Africa, long similarly used, is a more pungent and less aromatic substance. It used to be employed with ginger and cinnamon in making the spiced wine known as Hippocras, but is now used in fiery cordials and in cattle medicines.§

*Alpinia officinarum*, Hance, of the island of Hainan, has only been known as the source of the aromatic

* 'Pharmacographia,' pp. 540-5. Bentley and Trimen, iv. pl. 252.
† *Ibid.*, iv, pl. 270.
‡ *Ibid.*, iv, pl. 267. ' Pharmacographia,' pp. 582-9.
§ *Ibid.*, pp. 590-592 ; W. J. Hooker, ' Pharm. Journ.,' xii (1852), p. 192. Bentley and Trimen, iv, pl. 268.

rhizome GALANGAL since 1870.* It is used chiefly in Russia.

## IRIDACEÆ.

ORRIS-ROOT is the dried rhizome of *Iris Germanica*, L., *I. pallida*, Lam., and *I. Florentina*, L., long used in perfumery and in tooth-powder, having the odour of violets. It is imported from Leghorn, Trieste and Mogador.†

SAFFRON, used in pharmacy only as a colouring agent, but still used in Cornwall and elsewhere for colouring cakes, consists of the stigmas of *Crocus sativus*, L., a species unknown in a wild state. Its cultivation in Essex, which gave a name to the town of Saffron Walden, died out before 1768; but in Cambridgeshire it seems to have lingered into the present century. It is now grown in Spain and in the French department of Loiret.‡

## PALMACEÆ.

*Areca Catechu*, L., the ARECA-NUT, was added to the Pharmacopœia, as a tænifuge, in 1874. The Areca Palm is a native of the East Indies, where its small pear-shaped seeds are largely chewed with lime and the leaves of the Betel Pepper (*Piper Betle*, L.), and are therefore known as Betel Nuts. Its charcoal is used as tooth-powder.§

*Dæmonorops Draco*, Martius, a Malayan Rattan Palm, exudes from its fruits the dark-red resin known

* 'Journ. Linn. Soc.' (Botany), xiii (1873), p. 1 ; 'Pharmacographia,' pp. 580-582.   Bentley and Trimen, iv, pl. 271.
† *Ibid.*, iv, pl. 273.
‡ *Ibid.*, pl. 274.
§ *Ibid.*, pl. 276.

as DRAGON'S BLOOD, used for colouring plasters, tooth-powders, and varnishes.

### AROIDEÆ.

*Rhaphidophora*, sp., probably *R. vitiensis*, Schott, produces the root and stems which, together with the inactive leaves of the Verbenaceous *Premna taitensis*, Schauer, form the bulk of the native remedy known as TONGA, which has been found very efficacious in neuralgia.*

*Acorus Calamus*, L., the SWEET SEDGE, a naturalized British plant, yields an aromatic stimulant and tonic bitter from its rhizome, used in herb-beers, gin, and snuff, for chewing to clear the voice, for ague and dyspepsia. It is imported from South Russia, *viâ* Germany.†

### LILIACEÆ.

*Urginea Scilla*, Steinheil (= *U. maritima*, Baker, = *Scilla maritima*, L.), a Mediterranean species, yields the bulb known as SQUILLS, imported from Malta for use as an expectorant and diuretic. In larger doses it is an uncertain emetic and purgative.‡

*Aloe vulgaris*, Lamarck, the COMMON or BARBADOS ALOE, which grows wild from India to the South of Spain and the Canaries, and is cultivated in the West Indies ; *A. succotrina*, Lam., the SOCOTRINE ALOE, imported *viâ* Zanzibar and Bombay ; and *A. spicata*, L. fil., the CAPE ALOE, and other varieties,

* 'Pharm. Journ.,' x (1880), p. 889.
† 'Pharmacographia,' p. 613 ; Bentley and Trimen, iv, pl. 279.
‡ *Ibid.*, p. 627 ; *ibid.*, iv, pl. 281.

yield a valuable purgative, which is obtained as a bitter juice from the fleshy leaves. We import over 300 tons of Aloes annually.* Aloes have since 1869 formed an article of export from Natal.†

*Xanthorrhœa arborea*, R. Br., and *X. quadrangulata*, F. von Muell., the singular GRASS GUM TREES of Australia, contain abundance of Picric acid, used as a dye and in the preparation of the explosive potassium-salt.‡

### MELANTHACEÆ.

*Veratrum album*, L., WHITE HELLEBORE, a native of the mountains of Central and Southern Europe, and *V. viride*, Aiton, the INDIAN POKE, of North America, contain various alkaloids in their rhizome, which render them purgative, the former drastically so. They are mainly employed in veterinary medicine and in the United States.§

*Asagræa officinalis*, Lindl. (= *Schœnocaulon*, A. Gray = *Sabadilla*, Brandt), SABADILLA, a native of Mexico, contains Veratria, $C_{52}H_{86}N_2O_{15}$, useful as an ointment in rheumatism and neuralgia, in its seeds.‖

*Colchicum autumnale*, L., the MEADOW SAFFRON or AUTUMNAL CROCUS, the resemblance of which to the true Saffron Crocus is merely superficial, is a native of British limestone pastures, which has long

* 'Pharmacographia,' pp. 616-27 ; Bentley and Trimen, iv, pl. 282-4.
† Spon, ' Encyl. Indust. Arts.'
‡ Smith, ' Domestic Botany,' p. 160.
§ 'Pharmacographia, pp. 630-3 ; Bentley and Trimen, iv, pl. 285-6.
‖ *Ibid.*, p. 633; *ibid.*, iv, pl. 287.

been used in gout, rheumatism, and skin diseases, both its corms and seeds containing a principle named Colchicin. It is collected principally in Gloucestershire and in Germany.*

## SMILACEÆ.

*Smilax officinalis,* H. B. K., and *S. medica,* Schlecht. and Cham., woody climbers of New Granada and of Mexico respectively, are the main sources of SARSAPARILLA, the adventitious rootlets of which have for three centuries been largely used as an alterative and tonic. We import some 150 tons annually, the mealy or starchy varieties coming from Honduras, Guatemala, and to a small extent from Pará, and the non-mealy varieties from Jamaica, Mexico, and Guayaquil.†

## NAIADACEÆ.

*Elodea canadensis,* Rich., which, since 1840, has been looked upon as an unmitigated nuisance, is now said to be a germicide that checks malaria and epidemic diarrhœa.‡

## GRAMINACEÆ.

Various substances obtained from the Grass tribe appear in Pharmacopœias, such as Sugar, Pearl Barley, Rice-water, Groats, Flour, etc., but are merely vehicles or foods, not active drugs.

*Andropogon Nardus,* L., cultivated in Ceylon and

* 'Pharmacographia,' pp. 636-9; Bentley and Trimen, iv, pl. 288.
† *Ibid.,* pp. 639-47 ; *ibid.,* iv, pl. 289-90.
‡ Christy, ' New Commercial Plants,' No. 10, p. 37.

Singapore, yields CITRONELLA OIL ; *A. citratus,* DC.,
LEMON GRASS OIL, or OIL OF VERBENA, grown
throughout India, and *A. Schœnanthus,* L., non Wallich,
OIL OF GERANIUM, produced in Northern and Central India. These essential oils, used medicinally in
India, are imported to a considerable extent as per-
fumes, the latter being a frequent adulterant of Attar
of Rose.*

CONIFERÆ.

*Juniperus Sabina,* L., SAVIN, a native of Southern
Europe, contains in its young green shoots an oil
resembling turpentine, which is a powerful uterine
stimulant. The berries of the common JUNIPER (*J.
communis,* L.), a British shrub, are reputed diuretic.†

The LARCH (*Larix europæa,* DC.) yields VENICE
TURPENTINE, formerly used in veterinary medicine,
and its astringent bark is also used in bronchitis or
hæmorrhage.‡

*Tsuga Canadensis,* Carrière, the HEMLOCK SPRUCE
of Canada, yields the slightly stimulant CANADA
PITCH,§ used instead of BURGUNDY PITCH, the produce of the NORWAY SPRUCE, *Picea excelsa,* Link
(=*Pinus Abies,* L., *Abies excelsa,* DC.).||

*Abies balsamea,*¶ Mill., yields CANADA BALSAM, a
turpentine more used as a cement than as a drug ;

---

* 'Pharmacographia,' p. 660 ; Bentley and Trimen, iv., pl.
297.
† *Ibid.,* pp. 565-8 ; *ibid.,* iv, pl. 254-5.
‡ *Ibid.,* p. 551 ; *ibid.,* iv, pl. 260.
§ Bentley and Trimen, iv, pl. 264.
|| *Ibid.,* iv, pl. 261 ; 'Pharmacographia,' p. 556.
¶ *Ibid.,* iv, pl. 263.

and *A. pectinata,*\* DC. (=*Pinus Abies,* Du Roi =*P. Picea,* L.), the SILVER FIR, yields STRASSBURG TUR-PENTINE, more fragrant than common turpentine, but now seldom used.

COMMON TURPENTINE, in Europe, is produced by the SCOTS FIR, *Pinus sylvestris,*† L., the CLUSTER or MARITIME PINE, of Bordeaux, *P. Pinaster,*‡ Aiton, and the CORSICAN PINE, *P. Laricio,* Poiret ; and in America by the PITCH PINE, *P. Australis,*§ Michaux, and the LOBLOLLY, *P. Tæda,* L.||

*Pinus Laricio,* Poir., the CORSICAN PINE, is the species mainly employed in the manufacture of PINE-WOOL and FIR OIL, recently recommended for rheumatic and catarrhal affections.¶ The latter, when inhaled, affords relief to colds in the head.

## FILICES.

*Aspidium Filix-mas,* Swartz, has, since 1851, been administered as an ethereal extract. It is used chiefly as a tænifuge.\*\*

## FUNGI.

*Agaricus muscarius,* L., the FLY AGARIC, having been found to contain an alkaloid, MUSCARINE, wholly antagonistic in its action to Atropine, has recently attracted a good deal of attention in physiology ; but it is difficult altogether to account for the large ex-

---

\* Bentley and Trimen, pl. 262.
† *Ibid.,* pl. 257.               ‡ *Ibid.,* pl. 256.
§ *Ibid.,* pl. 258.               || *Ibid.,* pl. 259.
¶ S. Beaver, 'Pharm. Journ.,' iv (1863), pp. 424-5.
\*\* Bentley and Trimen, iv, pl. 300 ; 'Pharmacographia, p. 667.

portation of this fungus from Archangel. Other fungal poisons being apparently either Muscarine or some closely allied substance, Atropine may prove valuable as an antidote in the not infrequent cases of fungal poisoning.

*Claviceps purpurea*, Tulasne, the ERGOT OF RYE, belonging to the order Pyrenomycetes, is a fungus which occurs on many grasses, but is chiefly imported on Rye (*Secale cereale*, L.) from Vigo, in Spain, Hamburg, and Odessa. It is a violet-coloured, spindle-shaped 'sclerotium,' or compact mass of fungoid 'mycelium,' or 'spawn,' from ½ inch to 1½ inches long, entirely replacing the grain. It is known officinally as 'Secale cornutum,' and is employed to produce muscular contraction of the uterus.[*]

### LICHENS.

Of this group, now generally considered to consist of fungi living upon lowly-organized algæ in a state of 'symbiosis,' if not of parasitism, only one species, *Cetraria Islandica*, Acharius, ICELAND MOSS, has any claim to be considered medicinal. Though a native of our British hills, this lichen is mainly collected on barren hills in Sweden, and the mucilaginous modification of starch, known as lichenin, which it contains, renders it demulcent as well as nutritious.[†]

### ALGÆ.

Among members of this class, reference has already

[*] 'Pharmacographia,' p. 672 ; Bentley and Trimen, iv, pl. 303, and *supra*, p. 33.
[†] Prof. A. H. Church, 'Fooɑ,' p. 103 ; 'Pharmacographia,' p. 670 ; Bentley and Trimen, iv, pl. 302.

been made (pp. 49, 50, *supra*) to *Chondrus crispus,*
Lyngbye,* IRISH MOSS, and *Sphærococcus lichenoides,*
Ag., CEYLON MOSS,† which are used as demulcent
foods for invalids.

*Fucus vesiculosus,* L., the BLADDER-WRACK, the
commonest sea-weed on our coasts, besides being an
excellent manure, of use in times of scarcity as a cattle-
food, and when burnt into 'kelp,' a main source of
iodine, has been employed, when charred, as VEGE-
TABLE ETHIOPS, or as a jelly, in scrofulous tumours
and glandular enlargements, and is the essential consti-
tuent in the remedy for obesity known as ' Anti-fat.'‡

---

## PART III.—OIL-YIELDING SEEDS, VEGE-TABLE OIL, AND SUBSTANCES USED IN PERFUMERY.

THE term ' Oil' is difficult to define, and the diverse
substances included under it are accordingly difficult
to classify. Oils may be solid, viscid, limpid or volatile, odourless or odorous, and, though consisting
mainly of carbon and hydrogen, need not do so exclusively. They are mostly insoluble in water, and are
all readily inflammable. They may be physically
arranged under five groups, so far as they are of
vegetable origin :

  i. Non-drying, or greasy fluid oils.
  ii. Drying oils.

---

* Bentley and Trimen, iv, pl. 305.
† *Ibid.,* iv, pl. 306.          ‡ *Ibid.,* iv, pl. 304.

   iii. Fats.

   iv. Waxes.

   v. Volatile or essential oils.

Of these, the first four groups, known as the Fixed Oils, constitute a series of compounds formed, as was shown by the classic researches of the centenarian, M. Chevreul, between 1811 and 1823, of a glyceride, or body analogous to the alcohol glycerin, $C_3H_5(OH)_3$, in combination with a fatty acid having, in nearly all cases, the formula $C_nH_{2n}O_2$, or $C_nH_{2n-2}O_2$. Linoleic acid is $C_{16}H_{28}O_2$, and ricinoleic, $C_{18}H_{34}O_3$. The true waxes contain no glyceride; but some of those of vegetable origin do so. If heated with steam under pressure, the oil is decomposed into these two constituents: thus Stearin, $\left. \begin{array}{l} (C_{18}H_{35}O)_3 \\ C_3H_5 \end{array} \right\} O_3$, becomes Stearic acid, $\left. \begin{array}{l} 3C_{18}H_{35}O \\ \phantom{3}H \end{array} \right\} O$, and Glycerin. If treated with alkalies, or basic oxides in the presence of water, oils become saponified, the oxide uniting with the acid to form a soap, *i.e.*, generally a soda or potash salt of the acid, glycerin being liberated. This is the basis of the modern soap and candle industry, which, since the removal of the duty on soap, in 1852, has grown to an incredible extent.

The essential oils are freely soluble in alcohol, or in the fixed oils: they are not greasy, and at ordinary temperatures they are nearly all limpid liquids. They consist mainly of hydrocarbons, such as terpene, $C_{10}H_{16}$; but at a low temperature many of them separate into a liquid 'elæoptene' and a solid 'stearoptene.' Of these latter bodies, Camphor, $C_{10}H_{16}O$, may be taken as a type. Essential oils are used mainly in perumery for as flavouring in food.

The more important non-drying oils are the following :

| | | |
|---|---|---|
| Almond | from | *Amygdalus communis*, L. |
| Ben | „ | *Moringa pterygosperma*, Gaertn. |
| Brazil-nut | „ | *Bertholletia excelsa*, H. B. |
| Castor | „ | *Ricinus communis*, L.. |
| Colza | „ | { *Brassica campestris*, var. *oleifera*, DC. |
| Cotton-seed | „ | *Gossypium*, spp. |
| Croton | „ | *Croton Tiglium*, L. |
| Ground-nut | „ | *Arachis hypogæa*, L. |
| Mustard | | *Brassica*, spp. |
| Olive | „ | *Olea europea*, L. |
| Rape | „ | { *Brassica campestris*, var. *oleifera*, DC. |
| Sesame or Gingelly | „ | *Sesamum orientale*, DC. |

The chief drying oils are :

| | | |
|---|---|---|
| Candle-nut | from | *Aleurites triloba*, Forst. |
| German Sesame | „ | { *Camelina sativa*, Crantz ( = *Myagrum sativum*, L.). |
| Hemp | „ | *Cannabis sativa*, L. |
| Linseed | „ | *Linum usitatissimum*, L. |
| Madia | „ | *Madia sativa*, Willd. |
| Niger | „ | *Guizotia abyssinica*, Cass. |
| Nut | „ | *Juglans regia*, L. |
| Poppy | „ | *Papaver somniferum*, L. |
| Safflower | „ | *Carthamus tinctorius*, L. |
| Sunflower | „ | *Helianthus annuus*, L. |

9

The chief vegetable fats in commerce are :

Carapa, or
   Crab oil } from *Carapa guianensis*, Aub.

| | | |
|---|---|---|
| Cocoa-nut oil | „ | *Cocos nucifera*, L. |
| Cocoa butter | „ | *Theobroma Cacao*, L. |
| Chinese vege-<br>  table tallow } | „ | *Stillingia sebifera*, Michx. |
| Dika butter | „ | *Irvingia Barteri*, Hook. fil. |
| Kokum butter | „ | *Garcinia indica*, Choisy. |
| Mahwa butter | „ | *Bassia latifolia*, Roxb. |
| Shea butter | „ | *B. Parkii*, G. Don. |
| Nutmeg oils | „ | *Myristica moschata*, Thunb., etc. |
| Palm oil and<br>  Palm-seed oil } | „ | *Elais guineensis*, L. |
| Piney tallow | „ | *Vateria indica*, L. |

The vegetable-waxes, the first of which alone is of any commercial importance, are :

| | | |
|---|---|---|
| Japanese wax | from | *Rhus succedanea*, L. |
| Carnauba wax | „ | *Copernicia cerifera*, Mart. |
| Palm-tree wax | „ | *Ceroxylon andicola*, H. and B. |
| Myrtle-berry wax | „ | *Myrica cerifera*, L. |
| Ocuba wax | „ | *Myristica Ocuba*. |
| Cow-tree wax | „ | *Galactodendron*. |

Among the chief essential oils in commerce, to which reference will be made in the sequel, are the following :

| | | |
|---|---|---|
| *Eucalyptus* | from various species. | |
| Indian Geranium | „ | *Andropogon Schœnanthus*, L. |
| Lemon Grass | „ | *A. citratus*, DC. |
| Vetivert | „ | *A. muricatus*, Retz. |

Patchouli     from *Pogostemon Patchouli*, Pell.
Wintergreen     „   *Gaultheria procumbens*, L.

Of those enumerated above, Ground-nut, Brazil-nut, Olive, Poppy, Sunflower, Safflower, and nut oils, are used as food. Colza, Rape, Candle-nut, German Sesame, Madia, and Safflower are used as lamp-oils. In soap-making, the principal materials employed are Palm and Palm-kernel oil, Cotton-seed oil and Rosin, or Colophony, in ordinary soaps; Olive-oil in Castile and Brown Windsor; Cocoa-nut oil in hard soaps; Rape, Hemp, and Linseed oils in soft soaps; and Almond and Ground-nut oils and Cocoa-butter in various toilet-soaps. Various essential oils are added to give perfume, and many of the other oils mentioned are used to a less extent.*

In candle-making, besides such animal or mineral substances as Wax, Tallow, Spermaceti, Paraffin, and Ozokerit, the most important substance employed by far is PALM OIL, of which over 400,000 cwt. were imported, in excess of the like amount exported, in 1882. All the vegetable waxes, besides Palm-seed and Cocoa-nut oils, Dika, Shea, Mahwa, and Nutmeg butters and Chinese and Piney tallows are also used. Myrtle-berry wax was formerly used in the United States.

In perfumery, besides animal substances, such as ambergris, civet, musk, and the essential oils already referred to, use is made of many resinous substances, such as Myrrh, Storax, Benzoin, Opoponax, and the Balsams of Tolu and of Peru, and of a few artificially prepared arcmatic bodies. One of these latter,

Professor Church in Bevan's 'Manufacturing Industries.'

9—2

Vanillin, has been already mentioned. Another,
COUMARIN, is interesting from its occurrence in a great
variety of plants. Its composition is $C_9H_6O_2$, and it
has been found in *Dipteryx odorata,*Willd. (*Coumarouna
odorata,* Aubl.), the TONQUIN BEAN, used to scent
snuff, *Angræcum fragrans,* Reich. fil., the FAHAM TEA
of Mauritius, *Orchis militaris,* L., *O. Simia,* Lam., *O.
fusca,* Jacq., and *O. ustulata,* L., *Asperula odorata,* L.,
the SWEET WOOD-RUFF, once used to give perfume
to Rhine wines, *Eupatorium aromaticum,* L., and *E.
glutinosum,*Lam.,of North America, the Swiss *Melilotus
cærulea,* Desr., and our *M. officinalis,* Desr., and *Anthoxanthum odoratum,* L. Its presence in the latter gives
the fragrance to new-mown hay, and in the seeds of
FENUGREEK (*Trigonella Fœnum-græcum,* L.) it causes
their use to improve the appearance of inferior hay.

With these introductory remarks a few notes may
be added, taking the oils in systematic order.

### MAGNOLIACEÆ.

*Michelia Champaca,* L., yields an oil, which, with
that of *Cananga odorata,* Hook. fil. and Thom., and
that of the Cocoa-nut, is said to constitute the famous
MACASSAR OIL.*

### ANONACEÆ.

*Cananga odorata,* Hook. fil. and Thom., ILANG-
ILANG, a popular perfume, reached Europe first in
1864.†

* Guibourt, 'Histoire Naturelle des Drogues,' iii (1850), p.
675.
† 'Flora indica,' i, 130 ; Wiggers and Husemann, 'Jahresb.
d. Pharmac.' (1867), p. 422.

## PAPAVERACEÆ.

*Papaver somniferum*, L., MAW-SEED OIL, being very drying, is used by artists, or to adulterate Olive oil.

## CRUCIFERÆ.

*Brassica campestris*, var. *oleifera*, DC., RAPE or COLZA, yields from its seeds an oil still largely used for lamps and lubricating, though mineral oil has to a great extent superseded it. It is imported from France and Germany.

*B. juncea*, Hook. fil. and Thom., already alluded to (p. 63, *supra*), is extensively cultivated throughout India, and in other warm countries. Its seeds yield 20 per cent. of fixed oil, used in Russia instead of olive oil.

## BIXACEÆ.

*Gynocardia odorata*, R. Br., CHAULMUGRA, already mentioned.

## GUTTIFERÆ.

*Garcinia indica*, Choisy, from its seeds yields KO-KUM BUTTER, recommended for ointments, etc.

*Calophyllum inophyllum*, L., NDILO OIL, of Fiji, BITTER OIL, of India. Already referred to. Used for soap.

## DIPTEROCARPACEÆ.

*Vateria indica*, L., PINEY RESIN, INDIAN COPAL, or WHITE DAMMAR, prized for candle-making, from its pleasant smell, was not in use in 1851 [*]

[*] Prof. E. Solly in Society of Arts Lectures on the 1851 Exhibition.

## MALVACEÆ.

*Gossypium*, spp. COTTON-SEED OIL, though now one of the most important, has only been prepared since 1852, mainly from Egyptian cotton. 800,000 tons of seed (of which one quarter comes to England), yielding 120,000 tons of oil and 250,000 tons of the valuable oil-cake, are, it is estimated, manufactured annually. It is used for soap, lubricating oils and adulterating olive oil.

## LINACEÆ.

*Linum usitatissimum*, L., LINSEED, of which 2,433,132 quarters were imported in 1882, mainly as seed, from Russia and India, is the chief oil employed in painting (see p. 81, *supra*).

## GERANIACEÆ.

*Pelargonium capitatum*, Ait., is cultivated round the Mediterranean for its essential oil, known as OIL OF GERANIUM, which is used to adulterate Attar of Roses ;* but that obtained from *Andropogon Schœnanthus*, L., is known by the same name, and used for the same purpose.

## AURANTIACEÆ.

Reference has already been made to the essential OILS OF LEMON (*Citrus Limonum*, Risso, unripe fruit), BERGAMOT (*C. Bergamia*, var. *vulgaris*, Risso and Poiteau, unripe fruit), NEROLI, or ORANGE FLOWER (*C. vulgaris*, Risso, flowers), PETIT GRAIN or ORANGE LEAF (*C. vulgaris*, Risso, leaves and shoots), ORANGE

* ‘ Pharm. Journ.,’ ix (1852), p. 325 ; ‘ Kew Museum Guide,’ p. 24.

PEEL (*C. vulgaris*, Risso, unripe fruit), and CEDRAT (*C. medica*, Risso, fruit).*

## SIMARUBEÆ.

*Irvingia Barteri*, Hook. fil., DIKA, already mentioned (p. 60, *supra*), is used in soap and candle making. Containing myristine and laurine, it yields a very hard soap.†

## MELIACEÆ.

*Carapa guianensis*, Aubl., CARAPA or CRAB OIL, is said to be identical with *C. guineensis*, Juss., of West Africa, and is allied to *C. molluccensis*, L. It is one of the seeds known as Bitter Kola, and is used locally for rheumatism, but in Europe for soap, since 1851.‡ It is said to act as a preservative to wood.

## ANACARDIACEÆ.

*Rhus succedanea*, L., and *R. vernicifera*, DC., the LACQUER TREES of Japan, yield JAPANESE WAX, used for candle-making, from their fruits,§ now available to our commerce from the opening of Japanese ports.

## MORINGACEÆ.

*Moringa pterygosperma*, Gaert., and *M. aptera*, Gaert., yield BEN OIL, or OIL OF BEN, used as

* 'Pharmacographia,' pp. 106-16 ; Bentley and Trimen, i, pl. 50, 52, 53, 54.
† J. Attfield, 'Pharm. Journ.,' ii (1862), pp. 445-7 ; H. W. Batchelor, *ibid.*, xi (1880), p. 43.
‡ Prof. Solly, *op. cit.* ; Smith, 'Domestic Botany,' p. 462.
§ Hikorokuro Yoshida in 'Forestry and Forest Products,' Edinburgh, 1884.

salad oil in the West Indies, where it is cultivated, as a watchmaker's lubricant, and for hair oil.*

## LEGUMINOSÆ.

*Psoralea corylifolia*, L., BAWCHAN seeds of India, have been imported for oil-crushing.

*Arachis hypogæa*, L., GROUND-NUT, already mentioned (p. 39, *supra*), is the staple of the Marseilles soap-industry, the better qualities being substituted for olive. Three and a half million pounds are annually imported into France.

## ROSACEÆ.

ALMOND OIL, the glyceride of oleic acid, $C_{18}H_{34}O_2$, extracted as a fixed oil, by pressure, to the extent of about 45 per cent., from BITTER ALMONDS (*Amygdalus communis*, L., var. *amara*, DC.), or over 50 per cent. from SWEET ALMONDS (var. *dulcis*, DC.), and the essential BITTER ALMOND OIL, $C_7HO$, distilled from the residual cake, have been before mentioned (p. 59, *supra*).

ATTAR or OTTO OF ROSES, is a mixture of a liquid containing oxygen with an odourless solid hydrocarbon or stearoptene. It is obtained by distillation from the petals chiefly of *Rosa Damascena*, Miller, a variety of *R. gallica*, L., cultivated mainly on the lower slopes of the Balkans, in Eastern Roumelia, where about 4,000 lb., valued at £60,000, is annually produced. Little of that produced in South France, Tunis, Persia, or India reaches England, and Turkish

* Smith, *op. cit.*, p. 462 ; 'Encyclop. Brit.,' *sub voce* 'Oils'; 'Pharm. Journ.,' v (1845), p. 58.

Attar is largely adulterated with the Indian 'Geranium Oil,' obtained from *Andropogon Schœnanthus*, L.*

### MYRTACEÆ.

*Eucalyptus amygdalina*, Lab., *oleosa*, F. von Muell., *sideroxylon*, A. Cunn., *goniocalyx*, F. von Muell., *globulus*, Lab., *corymbosa*, Sm., *obliqua*, L'Hérit., *fissilis*, *odorata*, Behr., *rostrata*, Schlecht., *longifolia*, Link. and Otto, *viminalis*, Lab., *dumosa*, A. Cunn., and *citriodora*, Hook. all yield oils, which have been prepared by Mr. J. Bosisto, of Richmond, Victoria, some of which were known in 1851.

*E. Staigeriana*, F. von Muell., contains in its leaves 2 to 3 per cent., or 1,200 oz. to the ton, of an oil closely similar to Oil of Verbena (*Andropogon citratus*, DC.).

*E. hæmastoma*, Sm., yields one intermediate between Geranium and Peppermint, and it, and other Queensland species, seem well suited for scenting soaps.†

*Pimenta acris*, Wight., BAY-BERRY OIL, is used in the United States to flavour Bay Rum.

*Bertholletia excelsa*, H. B., the BRAZIL-NUT, yields the oil known as CASTANHA OIL, in Brazil, used by artists and watchmakers. Spoiled seeds are used for soap-making in Europe.

CAJEPUT OIL, used medicinally, which is distilled from the leaves of *Melaleuca leucadendron*, L., var. *minor*, has been before mentioned.

---

* 'Pharmacographia,' p. 233; Bentley and Trimen, ii, pl. 105.
† 'Pharm. Journ.,' xvii (1886); Christy, 'New Commercial Plants,' No. 9, p. 14.

OIL OF CLOVES, distilled to the extent of 16 to 18 per cent. from CLOVES, *Eugenia caryophyllata*, Thunb. (see pp. 66 and 94, *supra*), is extensively used by soap-makers and perfumers.

Sufficient reference has already been made to the volatile oil of DAMIANA (p. 95, *supra*), in the order TURNERACEÆ, and to those of CARAWAY, FENNEL, DILL, CORIANDER, and CUMIN (pp. 96, 97), among the UMBELLIFERÆ.

## COMPOSITÆ.

*Liatris odoratissima*, Willd., DEER'S TONGUE, or WILD VANILLA, has leaves used in perfumery in North America.*

*Guizotia abyssinica*, Cass., NIGER or RAMTIL seeds, came into the English market about 1851. It is a native of Tropical Africa, but is cultivated in India and Germany. It is used in Europe for soap and lubricating oil.†

*Helianthus annuus*, L., the SUNFLOWER, contains in its fruits a valuable oil, equal to Nut or Olive Oils, which is considerably used in Russia, and those of *Carthamus tinctorius*, L., SAFFLOWER, yield the KOOSUM OIL of India, and are used in soap-making.

## ERICACEÆ.

*Gaultheria procumbens*, L., of North-west America, contains in its leaves the volatile OIL OF WINTERGREEN.

---

* Bentley, 'Pharm. Journ.,' v (1874-5), pp. 489, 793.
† 'Encyclop. Brit.,' vol. xvii, p. 746, *sub voce* 'Oils,' by J. Paton.

## SAPOTACEÆ.

*Argania Sideroxylon,* R. and S., of Morocco, yields from its kernels an oil, known as ARGAN OIL, resembling Olive Oil.*

*Bassia latifolia,* Roxb., the MAHWA, besides its saccharine flowers, yields 33 per cent. of a butter from its seeds, used in India as food, and now imported for soap and candle making. It was not in commerce in 1851.† It is said to equal Cocoa-nut Oil.

*B. Parkii,* G. Don. (=*Butyrospermum,* Kotschy) yields from its kernels SHEA BUTTER, of which from 300 to 500 tons are exported annually from Sierra Leone, for soap-making. In the manufacture, from 5 to 75 per cent. of a hydrocarbon, 'Gutta Shea,' is extracted. This, also, is a recent introduction.‡

## OLEACEÆ.

*Olea europæa,* L., the OLIVE, long naturalized all round the Mediterranean, before mentioned (p. 60, *supra*), yields from its ripe fruits nearly 70 per cent. of the valuable non-drying OLIVE OIL. 23,450 tuns, of 252 gallons each, were imported in 1882. The finest, or 'Lucca Oil,' from Lucca, Florence, Leghorn and Genoa, is used in salads ; the less pure kinds, with soda, form Castile soap, and are used in dressing woollen, and for lubricating machinery.§

* D. Hay, 'Pharm. Journ.,' ix (1878), p. 262, and *ibid.,* x (1879), p. 127.
† Prof. Solly, *op. cit. ;* 'Encyclop. Brit.,' *loc. cit.*
‡ 'Kew Museum Guide,' and *op. sup. cit. ;* Smith, 'Domestic Botany,' p. 317.
§ 'Pharmacographia,' p. 374 ; Bentley and Trimen, iii., pl. 172.

PEDALINEÆ.

*Sesamum indicum,* DC., SESAME, TÎL SEED or GINGELLY OIL, is *the* Oil of India, and is used instead of, or as an adulterant of, Olive Oil, being itself adulterated with Ground-nut Oil. It is used here for soap-making ; but is chiefly crushed at Trieste and Marseilles. It contains 76 per cent. of Oleine, with stearic, palmitic and myristic acids.*

LABIATÆ.

Reference has already been made to the volatile oils and their contained camphors or stearoptenes of LAVENDER (*Lavandula vera,* DC.), SPIKE (*L. Spica,* DC.), MINT (*Mentha viridis,* L.), PEPPERMINT (*M. piperita,* Sm.), JAPAN PEPPERMINT (*M. arvensis,* L.), PENNYROYAL (*M. Pulegium,* L.), THYME (*Thymus vulgaris,* L.), and ROSEMARY (*Rosmarinus officinalis,* L.), which are used chiefly in perfumery.

*Lallemantia iberica,* Fisch. and Mey., one to three feet high, yields 2,500 seeds containing a pure oil suitable for food. It is cultivated in Syria, Persia, and South Russia.†

*Pogostemon Patchouli,* Pell., PATCHOULI, an essential oil, the perfume of which characterizes Indian shawls.‡

*Perilla ocimoides,* L., of Japan, yields from its seeds the YEGOMA oil used for Japanese leather-paper.§

* Spon's ' Encyclop. Indust. Arts ;' 'Pharmacographia,' p. 425; Bentley and Trimen, iii, pl. 198.
† Christy, 'New Commercial Plants,' No. 3, p. 14.
‡ 'Pharm. Journ.,' iv (1844), p. 80 ; vi (1847), p. 432 ; viii (1849), p. 574 ; ix (1850), p. 282 ; iv (1873), p. 362 ; xi (1880), pp. 409, 813.
§ ' Kew Museum Guide,' p. 105.

## MYRISTICEÆ.

*Myristica.*—Various species contain a good deal of a solid fat in their seeds, with more or less of the volatile oil of nutmeg. The following are imported : *M. angolensis*, Welw., 72 per cent. of oil ; no smell. *M. fatua*, Sw., of Borneo, 'LONG' or 'WILD' NUT-MEGS. *M. fragrans*, Houtt., of the Moluccas, the COMMON NUTMEG. *M. sebifera*, Sw., of Northern South America, VIROLA, yields 26 per cent. of AMERICAN NUTMEG OIL, used on the Continent for soap and candles. *M. surinamensis*, Rob., CUAGO nuts of the West Indies, imported in 1881, contain 65 per cent. of fat.*

## LAURINEÆ.

The essential OILS OF CINNAMON (*Cinnamomum zeylanicum*, Breyn.), CASSIA (*C. Cassia*, Bl.), and SASSAFRAS (*Sassafras officinale*, Nees), have already been referred to (see p. 113, *supra*).

*Persea gratissima*, Gaert., the AVOCADO PEAR, of Tropical America, has an edible berry which yields an oil suitable for soap-making or lighting purposes.

*Laurus nobilis*, L., the BAY, contains an aromatic oil, for which its leaves are employed in cookery ; whilst a perfume is also obtained from that of the berries.†

## EUPHORBIACEÆ.

*Aleurites triloba*, Forst. ( = *A. moluccana*, Willd. ?), of Fiji, etc., yields the CANDLE-NUT, COUNTRY

---

* Christy, No. 8, p. 26, with descriptions of other species and engravings of the seeds.
† Bentley and Trimen, iii, pl. 221.

WALNUT, KUKUI, KAKUNA or KĒKUNE OIL, exported from the Sandwich Isles, mainly to San Francisco. It is said to be equal to Rape Oil. It was imported from Fiji to the value of £3,040 in 1877.

*Stillingia sebifera*, Michaux, the CHINESE TALLOW-TREE, the seeds of which are coated with a fat mainly consisting of palmitin, now forms a regular import for soap and candle making.*

Reference has already been made to the medicinal CROTON OIL, obtained from *Croton Tiglium*, L., and to CASTOR OIL, from *Ricinus communis*, L., the inferior qualities of which are used for soap-making. That of the PHYSIC NUT (*Jatropha Curcas*, L.), a native of Tropical America, introduced into most tropical countries, is purgative, but has been imported for lighting and cloth-dressing, as SEED OIL.†

## JUGLANDACEÆ.

*Juglans regia*, L., the WALNUT, yields an edible oil, which, when expressed with heat, is a valuable drying oil. Similar oil is obtained from *J. cinerea*, L., the BUTTER-NUT, of the United States (see p. 58, *supra*).

## MYRICACEÆ.

*Myrica cerifera*, L., the BAYBERRY or WAX-MYRTLE of North America, *M. cordifolia*, L., of South Africa, and other species, secrete a wax on their berries, used in candle-making.

* 'Encyclop. Brit.,' *loc. cit.*
† Archer, 'Popular Economic Botany,' p. 265.

## CORYLACEÆ.

*Corylus Avellana,* L., the HAZEL, yields the valuable sweet oil already mentioned (p. 58, *supra*).

## QUERCINEÆ.

*Fagus sylvatica,* L., the BEECH, yields an oil from its 'mast,' or nut, which is used in France for food and for lighting.

## PALMACEÆ.

*Copernicia cerifera,* Mart., the WAX PALM of Brazil, secretes wax on its foliage, which is imported for candle-making.

*Cocos nucifera,* L., yields from its albumen the COCOA-NUT or COPRA OIL, of which 244,399 cwt. were imported for our soap and candle factories in 1884. It contains lauro-stearic, oleic, palmitic, myristic, and other acids, and yields a clear oil suitable for lamps.

*Elais guineensis,* L., yields PALM OIL from its mesocarp, and PALM-KERNEL OIL from its seed, the residue forming a useful cattle-food cake. Of 1,004,419 cwt. of Palm Oil imported in 1886, 925,000 cwt. came from West Africa, about half of that amount from British territory.*

## GRAMINEÆ.

Though it is unnecessary to refer again here to the essential oils upon which such substances as Carda-

* 'Pharm. Journ.' iv (1873), p. 531, and viii (1877), p. 68.

moms, Galangale and Calamus depend for their phar-
maceutic value, mention ought, perhaps, to be again
made of the Grass oils, perfumed volatile oils, distilled from fresh plants.

*Andropogon Nardus*, L., yields CITRONELLA OIL,
used in scenting Honey Soap.*

*A. citratus*, DC., LEMON GRASS, yields OIL OF
VERBENA or LEMON.   This was figured in Wallich's
' Plant. Asiat. Rar.' (t. 280) as *A. Schœnanthus.*

*A. Schœnanthus*, L., non Wallich (=*A. Martini,*
Roxb., *A. Pachnodes*, Trinius, *A. Calamus-aromaticus,*
Royle) yields the RÚSA OIL, OIL OF GINGER GRASS,
or OIL OF GERANIUM, used to adulterate Attar of
Roses.

### CONIFERÆ.

The hydrocarbon OILS OF SAVIN (*Juniperus Sabina,*
L.), JUNIPER (*J. communis*, L.), SPRUCE or HEMLOCK
(*Tsuga Canadensis*, Carr.), and TURPENTINE (*Pinus
sylvestris*, L., *P. Pinaster*, Sol., *P. Laricio*, Poir., *P.
longifolia*, Lambert, of the Himalayas, *P. Tœda*, L.,
and *P. australis*, Michx., *Abies pectinata*, DC., and
*Larix europœa*, DC.), have been already sufficiently
referred to (see pp. 124, 125, *supra*).

---

## PART IV.—GUMS, RESINS, OLEO-RESINS,
## AND INSPISSATED SAPS.

As in external form, so in internal and chemical con-
stitution, the variety of Nature and the minuteness

* Bentley and Trimen, iv, pl. 297 ; ' Pharmacographia,' p.
660.

of the graduations which she presents to our notice tax to the uttermost our powers of classification. This is well illustrated in the series of substances at present under consideration, for which no one comprehensive name, nor any entirely satisfactory scheme of classification, has as yet been suggested. The names 'Gum' and 'Balsam' have been very loosely employed in the past, and there is, in fact, an insensible gradation from the limpid essential oils to the solid resins. The essential oil known as Attar of Rose is itself solid at ordinary temperatures, and reference has already been made to the solid stearoptenes present in varying proportions in many others of that series. When the resin is so incompletely dissolved in the essential oil as to form a viscous body, it is termed an oleo-resin, and it has been proposed to confine the term 'balsam' to those fragrant substances that contain cinnamic, or some analogous acid, in addition to the volatile oil and resin of which the turpentines alone consist. 'Gums,' properly so called, are exudations from plants, soluble in water, at least in part, forming with it a mucilage insoluble in alcohol of 60 per cent., convertible by sulphuric acid into dextrine, and with nitric acid yielding mucic and oxalic acids. They are quite amorphous, being neither organized like starch, nor crystallizable like sugar. Resins, on the other hand, are insoluble in water, but mostly soluble in alcohol, essential oils, ether, or heated fatty oils; non-crystalline, incapable of sublimation, burning with a bright but smoky flame, containing little oxygen and no nitrogen. A typical resin is a pale yellow, transparent solid, with a glass-

like fracture and little or no smell or taste. The
term 'Gum-resin' is a correct designation for various
inspissated plant-saps which contain both gum and
resin.

These definitions necessarily exclude that distinct
class of substances derived from the latex of plants,
and apparently present to some extent in all plants,
the caoutchoucs, or, more conveniently, 'Rubbers.'
These are insoluble in water, alcohol, or unconcen-
trated acid, and are essentially hydro-carbons, probably mixtures of two or more, together with an
oxidized substance.   The introduction of Gutta
Percha in 1844, which will always redound to the
credit of the Society of Arts, and the discovery that
it; too, like Caoutchouc, might be made to unite
chemically with sulphur, thus becoming 'vulcanized'
—a discovery only made in the case of the other
substance a few years previously—are among the
most revolutionary events in the history of commerce.

It will be convenient to deal first with the true
Gums, then with the Resins, oleo-resins and gum-
resins, and, lastly, with the Caoutchoucs and Guttas.
The chief gum-yielding plants are the following :

### MIMOSEÆ.

*Acacia.* Various species, ·of which more will be
said, yielding the ARABIC, SENEGAL, and WATTLE-
GUMS, and part of the INDIAN TRAGACANTH or
KUTEERA.

*Prosopis dulcis*, Kunth., and other species, such as
*P. pubescens*, Benth., of Mexico and Texas, and

*P. juliflora*, DC., of Jamaica, yielding the valuable GUM MEZQUITE, now coming into use.*
*Parkia*, spp.

## PAPILIONACEÆ.

*Astragalus gummifer*, Labil., and other species of south-west Asia, yielding the true TRAGACANTH† or GUM DRAGON.

## DRUPACEÆ.

*Prunus* and *Amygdalus*, spp., yield CHERRY GUM, which is not used commercially.

## BIXACEÆ.

*Cochlospermum Gossypium*, DC., of India, the main source of KUTEERA.

## ANACARDIACEÆ.

*Anacardium occidentale*, L., CADJII.
*Spondias mangifera*, Pers., HOG-GUM.
*Rhus Metopium*, L.
*Odina Wodier*, Roxb., GING, KUNNEE, SHIMPTEU, MOOI. Growing in Travancore and the Coromandel Coast, and used in dyeing, cloth-printing, and Indian ink, worth 10s. to 30s. per cwt.; but containing a large percentage of insoluble matter, so, perhaps, only useful here to adulterate other gums. Nevertheless, as Dr. George Watt says :‡ 'The gums of India have

* 'Pharm. Journ.,' vi (1875), p. 942.
† P. J. Giraud, *ibid.*, 1878, No. 405 ; 'Pharmacographia,' p. 152.
‡ 'Catalogue of Indian Exhibits : Colonial and Indian Exhibition,' p. 137. Yet in 1886 we imported 33,581 cwt. of 'unenumerated gums' from India, thrice as much as from Turkey.

been entirely overlooked, and the result is that not a single Indian gum is to-day of any commercial value.'

### AURANTIACEÆ.

*Feronia elephantum*, Corr., WOOD-APPLE GUM, of South-west India. Cheap and useful.

*Ægle Marmelos*, Corr., said to yield 'a good arabic.'

### MELIACEÆ.

*Melia Azedarach*, L.

### BOMBACEÆ.

*Bombax Malabaricum*, DC., etc., MUCHERUS.
*Adansonia digitata*, L.

### STERCULIACEÆ.

*Sterculia urens*, Roxb., etc., of India, and *S. tragacantha*, Lindl., etc., of tropical Africa. KUTEERA or AFRICAN TRAGACANTH.

### CACTACEÆ.

*Cactus* and *Opuntia*, spp.

### MORINGACEÆ.

*Moringa pterygosperma*, Gaertn., a red gum, used in Indian medicine.

### BROMELIACEÆ.

*Puya*, spp. CHAGUAL, from S. Iago de Chili, resembles Gum Arabic, 75 per cent. soluble. It has only recently been introduced.

PALMACEÆ.

*Cocos nucifera,* L.

Of these, the only group calling for further remarks are the Acacia-gums. The chief gum-yielding species are the following :

*A. Senegal,* Willd. (=*A. verek,* G. and P.), yielding most of the best GUM ARABIC, PICKED TURKEY, WHITE SENAAR, or GUM SENEGAL, the best from Kordofan ; that from Senegal coming *viâ* Bordeaux.

*A. stenocarpa,* Hochst.    ⎱ SUAKIM or TALCA
*A. Segal,* Del., var. *Fistula* ⎰    GUM.

*A. gummifera,* Willd., MOROCCO, MOGADOR, or BROWN BARBARY, according to Messrs. Hooker and Ball.

*A. horrida,* Willd., CAPE.

*A. arabica,* Willd., BABUL or EAST INDIAN GUM ARABIC, from Africa, but coming *viâ* Aden and Bombay.

*A. Catechu,* Willd., KHEIR.

*A. pycnantha,* Benth., GOLDEN WATTLE, of Australia.

*A. dealbata,* Link., SILVER WATTLE, of Australia.

*A. decurrens,* Willd., BLACK or GREEN WATTLE.

*A. homalophylla,* A. Cunn., and perhaps other species.

CARAMANIA or BASSORA GUM, also known as HOG TRAGACANTH, and containing only one per cent. of soluble gum, is said to be derived from a *Prunus* or *Amygdalus.*

Following an anonymous writer in the 'Encyclopædia Britannica,' vol. xx., we may subdivide the Resins as follows :

1. COPALLINE or VARNISH RESINS, the purest, in-

cluding AFRICAN COPAL, or GUM ANIMÉ,* that of
ZANZIBAR (Bombay), from *Trachylobium Horneman-
ianum*, Hayne, as shown by Sir John Kirk ; that of
WEST AFRICA, from *Guibortia copallifera*, Bennett ;
that of SIBERIA, according to Mr. E. M. Holmes, from
a species of *Icica;* MEXICAN COPAL, JUTAHÝ-SEEA,
from *Hymenæa Courbaril*, L., etc.; BRAZILIAN COPAL,
from *Hymenæa* and *Trachylobium martianum*, Hayne ;
PINEY RESIN, or WHITE DAMMAR, or INDIAN COPAL,
from *Vateria indica*, L., and *V. acuminata*, Hayne, the
latter of Ceylon ; SAL DAMMAR, from *Shorea robusta*,
Gaertn., etc. ; DAMMAR, from *Hopea robusta*, Roxb.,
etc.; BLACK DAMMAR, from *Canarium strictum*, Roxb.,
of Malabar ; MASTIC, from *Pistacia Lentiscus*, L., and
BOMBAY MASTIC, from *P. cabulica*, Stocks, and *P.
Khinjuk*, Stocks ; LAC, produced by the punctures of
*Schleichera trijuga*, Willd., *Butea frondosa*, Roxb.,
*Ficus religiosa*, L., and many other trees rich in gum,
resin or saponaceous juice, by *Coccus Lacca*† ; EAST
INDIAN DAMMAR(*Dammara orientalis*,Lamb.); KAURI
or COWDIE GUM or RESIN (*D. australis*, Lamb.), of
New Zealand‡ ; produced also by *D. ovata*, Moore,
and *D. Cookii; D. lanceolata*, Vieill., of New Caledonia,
*D. robusta*, C. Moore, of Queensland. It is also found
fossil. Selected, it is worth from 115s. to 200s. per
cwt. It is, as is all Dammar, an introduction dating
mainly from 1851 ; but 5,500 tons were imported in
1880. *D. vitiensis*, Seem., yields FIJIAN COPAL.
SANDARAC, from *Callitris quadrivalvis*, Vent., of

* O. Crease, 'Pharm. Journ.,' vii (1847), p. 15.
† J. E. O'Conor, ' Lac,' Calcutta, 1876.
‡ ' Pharm. Journ.,' xi (1881), p. 939.

Algeria, and *C. verrucosa*, R. Br., *robusta*, R. Br., and *cupressiformis*, Vent., of Australia; and DRAGON'S BLOOD (*Dæmonorops Draco*, Mart.).*

2. SOFT or OLEO-RESINS, including MANILA ELEMI (*Canarium commune*, L.); MEXICAN ELEMI (*Amyris elemifera*, Royle), introduced about 1850; BRAZILIAN ELEMI (*Icica Icicariba*, DC., etc.); EAST INDIAN TACAMAHAC, produced by *Calophyllum inophyllum*, L., *C. Calaba*, L., *Icica Tacamahaca*, H., B., and K.; AMERICAN, by *Elaphrium tomentosum*, Jacq., and *Populus balsamifera*, L.; WOOD OIL or GURJUN BALSAM, from *Dipterocarpus alatus*, Roxb., and *D. turbinatus*, Gaertn., of Burma, etc.; CHIAN TURPENTINE (*Pistacia Terebinthus*, L.); TURPENTINE, Common FRANKINCENSE and THUS, from the various *Coniferæ*, and CANADA BALSAM (*Abies balsamea*, Mill.).

3. FRAGRANT OLEO-RESINS and GUM-RESINS, including MYRRH (*Balsamodendron Myrrha*, Nees); BDELLIUM or GOOGUL (*B. Roxburghii*, Arn., or *B. Mukul*, Hook.†); BALM OF GILEAD or MECCA BALSAM (*B. Opobalsamum*, Kunth, or *B. Berryi*); OLIBANUM or FRANKINCENSE (*Boswellia Carteri*, Birdw.); BENZOIN (*Styrax Benzoin*, Dry.); STORAX (*Styrax officinale*, L.), no longer obtainable; LIQUID STORAX (*Liquidambar orientalis*, Mill.); BALSAM OF PERU (*Myroxylon Pereiræ*, Klotzsch) and of TOLU (*M. Toluiferum*, H. B. K.) and LABDANUM (*Cistus creticus*, L., var. *labdaniferus*).

4. FETID GUM-RESINS, including AMMONIACUM (*Dorema Ammoniacum*, Don); ASAFŒTIDA (*Ferula*

* 'Pharmacographia,' p. 609.
† Journ. of Botany, 1849.

*Narthex*, Boiss., and *F. Scorodosma*, Benth.); GAL-
BANUM (*F. galbaniflua*, Boiss. and Buh., and *F.
rubricaulis*, Boiss.); OPOPONAX (*Opoponax Chironium,*
Koch), and SAGAPENUM (*Ferula* sp.). (See pp. 96, 97.)

5. MEDICINAL RESINS, including GAMBOGE, GUAIA-
CUM, EUPHORBIUM and COPAIBA ; to which may be
added DIKA MALI, from *Gardenia lucida*, Roxb. ; and

6. EXTRACT RESINS, including SCAMMONY, JALAP,
PODOPHYLLUM, CHURRUS (from Hemp) and CUBEBS.

The Natural Orders yielding these resins are :

LEGUMINOSÆ: *Myrospermum, Copaifera, Trachylobium, Hymenæa.*

ANACARDIACEÆ : *Pistacia, Rhus.*

AMYRIDACEÆ : *Boswellia, Icica, Bursera, Canarium, Amyris, Balsamodendron, Elaphrium.*

ZYGOPHYLLACEÆ : *Guaiacum.*

EUPHORBIACEÆ : *Euphorbia.*

DIPTEROCARPACEÆ : *Vateria, Shorea, Hopea, Dryobalanops, Dipterocarpus, Vatica.*

GUTTIFERÆ : *Garcinia, Calophyllum.*

CISTACEÆ : *Cistus.*

UMBELLIFERÆ : *Ferula, Opoponax, Dorema.*

RUBIACEÆ : *Gardenia.*

CONVOLVULACEÆ : *Convolvulus, Ipomœa.*

STYRACACEÆ : *Styrax.*

MORACEÆ : *Ficus.*

ALTINGIACEÆ : *Liquidambar.*

LILIACEÆ: *Xanthorrhœa*, yielding Gum Acaroides,
Botany Bay or Black-boy Resin, first introduced about
1850.*

---

* P. L. Simmonds, 'American Journ. of Pharm.,' 1857, p. 226 ;
1866, p. 465.

PALMACEÆ : *Dæmonorops.*

CONIFERÆ : *Abies, Pinus, Dammara, Callitris.*

In this classification no place has been found for the CAMPHORS, which are, as has been pointed out, the stearoptenes of essential oils rather than true resins. Of CAMPHOR there are, commercially speaking, three kinds :

(i.) COMMON or LAUREL CAMPHOR, from *Camphora officinarum*, Nees. (=*Laurus*, or *Cinnamomum, Camphora*), of China and Japan, grown mainly in Formosa, which, since 1868, has been free to European trade.

(ii.) BORNEO or MALAY CAMPHOR or KAPUR BARUS, shipped from Barus, sometimes called BAMBOO CAMPHOR, from its mode of packing, from species of *Dryobalanops.*

(iii.) NGAI CAMPHOR, of China, unknown in Europe, though £3,000 worth is annually exported from Canton. It is derived from *Blumea grandis*, DC., a tall weed of the Tenasserim provinces, and from *B. balsamifera*, DC., and is used in the ink-manufacture.*

Though INDIA-RUBBER and other forms of CAOUTCHOUC have long been known, the Caoutchouc trade may truly be said to be that of the last quarter of a century. For in 1830 we only imported 464 cwt., in 1840, 6,640 cwt., in 1850, 7,616 cwt., but in 1870, 152,118 cwt., and in 1886, 194,748 cwt. Nearly half of this last amount, which is exclusive of GUTTA PERCHA, came from Brazil, the Gold Coast and Bombay contributing the next largest amounts. The enormous extension of the trade between 1850 and

* Bentley and Trimen, iii, pl. 222, ' Pharmacographia,' pp. 458-466.

1870 was no doubt owing to the general adoption of the process of vulcanizing. Until the introduction, a few years ago, of the remarkably cheap and protean material known as CELLULOID, the useful, and even necessary, appliances to which Rubber or Vulcanite alone was applicable seemed practically infinite. Naturally the Tropics, where alone the trees seem to produce the abundant latex required, have been ransacked for new Rubber-yielding species. These seem to grow mainly between the isotherms of 70° F., where the annual rainfall is about 90 inches. The chief kinds in commerce are American :—Pará, Carthagena (from New Granada), Ceará ; West Indian :—Guayaquil, from Ecuador, Pernambuco, Maranham, Nicaragua, Honduras, and Guatemala; Asiatic :—Singapore, Assam, Penang, Java, Siam and Borneo ; and African : —Madagascar and West Coast. Of these, the PARÁ, or CAHOUT-CHOU, from the Amazon valley, is the most valuable. It is the produce of *Hevea guianensis*, Aub., *brasiliensis*, Müll. Arg., *Spruceana*, Müll. Arg., *paucifolia*, Müll. Arg., *rigidifolia*, Müll. Arg., and *discolor*, Müll. Arg. CEARÁ RUBBER is derived from *Manihot Glaziovii*, Muell. Arg. ; that of Pernambuco from *Hancornia speciosa*, Gomes, the MANGABEIRA RUBBER-TREE. The WEST INDIAN RUBBER comes from the mainland, and is the produce (as is that of Carthagena, Guayaquil, Nicaragua, Honduras and Guatemala, known as the ' Ulé group ') of *Castilloa elastica*, Cerv., and *C. Markhamiana*, Collins, the latter discovered in 1872. The caoutchoucs of Assam or Sylhet, and probably Singapore, are derived from *Ficus elastica*, Roxb., *F. laccifera*, Roxb., and *F. hispida*, Roxb., and

that of Borneo, GUTTA SUSSU, from *Willughbeia*
(= *Urceola*) *elastica*, discovered in 1864. GUTTA
SINGGARIP, of Malaysia, comes from *W. firma*, Bl.,
and *Leuconotis eugenifolius*, A. DC. also contributes to
the rubber of Borneo and Perak. African rubber is
mainly procured from species of *Landolphia*, which,
with those of other rubber-yielding genera, have been
much elucidated by the researches of Mr. W. T. T.
Dyer, Director of the Royal Gardens, Kew. *L. florida*,
Benth., furnishes MBUNGU RUBBER from the East
Coast, occurring also on the West ; *L. Kirkii*, Dyer
(MATERE RUBBER), and *L. Petersiana*, Dyer, from the
East Coast, and *L. owariensis*, Beauv., from Central
and West Africa, being the chief remaining species.
*L. florida* occurs in Angola, and *L. Heudelotii* in Senegal.
*Ficus Brasii*, R. Br., also yields rubber at Sierra Leone.
Various species of *Vahea*, *V. Madagascariensis*, VOÂ-
HÉRÉ or VOÂ-CANJA, *V. Comorensis*, VOÂ-HINÉ, *V.
gummifera*, etc., contribute the rubber of the Mozam-
bique coast, Madagascar and the Comoro Islands,
second only in quality to that of Pará. In 1875 Mr.
Robert Cross, who assisted so materially in the introduction of *Cinchona* into India, obtained plants of
*Castilloa elastica*, and in 1876 some of *Hevea brasiliensis* for Ceylon and India, where there is every
prospect of their cultivation succeeding.

GUTTA PERCHA is obtained from various Sapota-
ceous plants in the Malay Archipelago, its name being
perhaps more correctly GUTTA TABAN, Percha being
the name of an island where it is obtained. In 1822,
Dr. William Montgomerie noticed the substance in
use at Singapore, and, being again stationed there in

1842, sent home specimens and gave a lecture on the subject before the Society of Arts, whose gold medal he received. The trees from which it is obtained range from 6° or 10° N. to 10° S. lat., and from 100° to 120° E. long., and, being more easily so used, have been felled in great numbers since 1845, ten trees only yielding about 130 lb. of gutta. This was the amount sent over in 1844. In 1857, 4½ million lb., in 1876, 10 million, and in 1882 over 8 million lb. were imported, whilst the slackening of the supply has already considerably enhanced the price. In 1881, G. E. C. Beauvisage, published, at Paris, his ' Origines Botaniques de la Gutta Percha,' in which he refers to *Dichopsis Gutta*, Benth., the ' taban ' of the Rion Archipelago, ' percha,' ' derian,' ' dadu,' ' seroja,' ' tambaga,' and ' balam,' to *Isonandra macrophylla*, De Vriese, the second quality, ' putih,' to *I. Motleyana*, De Vriese, the inferior ' kotian,' and those of various other kinds from the Banka Islands to other species of *Isonandra*. The latest botanical account of the plants yielding Gutta Percha, however, is that of Burck.* He states that *Palaquium Gutta*, Burck. ( = *Dichopsis*) no longer exists in a wild state, that most exported now being ' taban simpor,' *Payena Maingayi*, while the best quality is ' taban marah,' *Palaquium oblongifolium*, Burck. He re-describes all the forms, and suggests as suitable *P. Borneense*, Burck, *P. Treubii*, Burck, *P. Treubii* var. *parvifolium*, Burck, and *Payena Leerii*, Burck, to which may be added ' Gutta jelutong,' *Dyera laxiflora*, Hook, fil., from Selangor.†

* Ann. du jard. bot. Buitenzorg, v (1886), p. 1.   ' Pharm Journ.,' xvii (1886).        † ' Kew Report,' 1881.

Among substitutes for Gutta Percha, the most important is BALATA (*Mimusops globosa*, Gaertn.), introduced from British Guiana in 1859.* Another likely to prove valuable is PAUCONTEE (*Bassia elliptica*, Dalz.), from the West Coast of India,† whilst MUDAR (*Calotropis gigantea*, R. Br.), *Euphorbia Cattimandoo*, Elliott, and other species of the same genus, from the same country, will no doubt become important.

## PART V.—DYES AND TANNING MATERIALS.

IN no class of vegetable products has chemistry so largely superseded the compounds prepared by Nature, as it has in the case of those used in dyeing and tanning. The invention of the great and ever-lengthening series of aniline colours, and the preparation, in 1869, of Alizarine, the colouring-matter of Madder, by synthesis, and now, again, the introduction of the chrome-process in tanning, have had a most striking, and, in some respects, not altogether a satisfactory effect. Dr. George Watt, for instance, says‡ that 'these cheap colours have not only depraved the tastes of the people, but have demoralized their indigenous industries;' much less Safflower, Lac, Indigo and Cochineal being exported now than ten years ago, whilst even Manjit, Saffron, and Myro-

* G. S. Jenman, 'Timehri,' iv, p. 219.
† Hugh Cleghorn, 'Paucontee,' Madras, 1858, 4to.
‡ 'Catalogue of Indian Exhibits : Colonial and Indian Exhibition,' p. 148.

balans are imported from Persia, France, and Ceylon respectively, instead of being produced at home. The knowledge of indigenous dyes possessed by the people is threatened, and turmeric, possibly from its use for other purposes, is 'the only dye-stuff which seems to have prospered, in spite of the introduction of aniline colours.' In Europe and the Levant, artificial alizarine has almost extinguished the cultivation of Madder; but, nevertheless, some new dyes of botanical interest have been introduced during the last half-century, whilst the general extension of trade, consequent on the mere growth of our population, has led not only to increased importation of the astringent substances used both in dyeing and tanning, but also to the introduction of novel substances for these purposes. Among the former (new dyes) the most interesting are the Indian MANJIT or MUNJEET (*Rubia cordifolia*, L.), which is, however, but rarely brought over, 525 tons arriving in 1850, but none in 1851 or 1852; the CHINESE GREEN INDIGO or LO-KAO, from the barks of *Rhamnus chlorophorus*, Dev., and *R. utilis*, Dev., which, Mr. W. B. Hemsley states, are identical with *R. tinctorius*, W. K., and *R. dahuricus*, Pall., respectively, which has been superseded by the less permanent aniline dyes, though much praised as a novelty thirty years ago;* and TOKIO PURPLE, the Japanese *Lithospermum erythrorhizon*, which may prove of use for colouring oils. Indigo and Orchil have been prepared chemically, but not as yet in commercial quantities.

As a tanning material, English Oak-bark stood

* Persoz., 'Comptes Rendus,' xxxv, 558.

almost alone; but while the consumption of this material has rather diminished, other substances have been imported in continuously increased quantities. One of the latest full accounts of these is that given by Mr. Thomas Christy, 'New Commercial Plants and Drugs,' No. 5 (1882), in which he classifies them as barks, fruits, galls, leaves, roots, and extracts.

MIMOSA-BARK, or WATTLE, derived from *Acacia mollissima*, Willd., *decurrens*, Willd., *pycnantha*, Benth., and *dealbata*, Link., has been an article of import from Australia and Tasmania for many years. The last-mentioned is very inferior ; *A. decurrens*, var. *mollis* (= *A. mollissima*, Willd.), is the best. An extract, or catechu, is prepared from the latter. *A. retinodes*, Schlecht., is also valuable, as is also the BABUL BARK (*Acacia arabica*, Willd.) of India, which is as yet hardly known in England.

MANGROVE BARK (*Rhizophora Mangle*, L.) gives a bad colour and quality to the leather if used alone, but might prove useful if mixed with other materials.

RED QUEBRACHO (*Loxopterygium Lorentzii*, Griseb.) is a hard wood from the River Plate, recently introduced, difficult to use from its great density. Other Quebrachos, viz., blanco ' (*Aspidosperma Quebracho*), flojo' (*Iodina rhombifolia*, Hook and Arn.), and ' tipa ' (*Machærium fertile*, or *Tipuana speciosa*, Benth.), are also used in South America, and 200 tons were imported in 1880, at £4 10s. per ton.

BALSAMOCARPON, the fruit-pod of *Cæsalpinia brevifolia*, the ALGARROBO of Chili, was introduced in 1876 ; but probably two or more species of *Prosopis* are included under this name or ALGAROBILLA, *P.*

*pallida,* Kunth, being 'Algarobilla negro,' and *P. Algarobo* being 'Algarobilla blanco.' 50 tons were imported in 1880, 160 in 1881 (H. R. Proctor, 'Text-book of Tanning,' p. 23).

DIVI-DIVI, the pods of *Cæsalpinia coriaria,* Willd., of Central America, has been longer in general use.*

BETEL-NUT (*Areca Catechu,* L.) is recommended by Mr. Christy.

MYROBALANS were only introduced about 1840, but we now import seven or eight thousand tons from India annually. The four varieties are the Chebulic (*Terminalia Chebula,* Retz.), Citrine (*T. citrina,* Roxb.), Beleric or Bastard (*T. belerica,* Roxb.), and the Emblic (*Emblica officinalis* Gaertn.). They yield valuable black and yellow dyes.

VALONIA, the 'cups' of *Quercus Ægilops,* L., together with 'Cameta' and 'Cametina,' its young acorns, have long been imported from the Levant. In addition to the long-known galls of *Q. infectoria,* Oliv., in 1847 the first importations were made of those produced by *Aphis chinensis* on *Rhus succedanea,* L., and *R. semialata,* Murray, of Northern India, China, and Japan, known as 'Wu-pu-tze' or Chinese or Japanese galls, of which we imported, mainly for exportation, in 1880, 51,083 cwt. at 50s. to 70s. per cwt. They yield 70 per cent. of tannin.

Mr. Christy also recommends the Galls of *Tamarix.*

SUMACH, the leaves and twigs of *Rhus Coriaria,* L., *R. typhina,* L., and (Venetian) *R. Cotinus,* L., is an old and valued material.

* 'Pharm. Journ.,' v (1846), p. 443.

CANAIGRE is the root of the Texan Dock (*Rumex hymenosepalus*, Torrey), recommended by Mr. Christy.

MIMOSA EXTRACT was sent from Australia as early as 1823, but dropped out of notice till recently. In 1880 we imported £682,296 worth of various Bark-extracts.

TERRA JAPONICA or GAMBIR, extract of *Uncaria gambir*, Roxb., and *U. acida*, Hunt., of Singapore, of which 3,234 cwt. were imported in 1830, we received to the extent of over 23,000 tons in 1875.

HEMLOCK EXTRACT, from *Tsuga Canadensis*, Carr., was first patented in 1864. 190,000 dollars' worth was made in 1881, 305,000 dollars' worth in 1883.

CHESTNUT EXTRACT, from *Castanea sativa*, Scp., has been recently imported from the Continent; and one of the latest introductions, a result of the Edinburgh Forestry Exhibition of 1884, is the bark of the Sal of India (*Shorea robusta*, Gaertn.), which yields 32 per cent. of tannin.*

KINOS are too costly for tanning: EAST INDIAN is obtained from the wood of *Pterocarpus marsupium*, Roxb.; AFRICAN from that of *P. erinaceus*, Poiret; BENGAL, PULAS, or DHAK, from *Butea frondosa*, Roxb.; WEST INDIAN from *Coccoloba uvifera*, Jacq., and AUSTRALIAN from *Eucalyptus resinifera*, Sm., etc.

Among dye-stuffs that have long been in use are the following:

LOGWOOD (*Hæmatoxylon campechianum*, L.), a Leguminous tree, native of Central America, introduced into Jamaica in 1815, of which the heart-wood is employed as a red or black dye. Its price varies

* 'Leather, July, 1886.

from £5 to nearly £10 per ton, and we import from 40,000 to 60,000 tons annually.*

*Cæsalpinia echinata*, Lam., BRESIL DE ST. MARTHA, an allied tree, yields NICARAGUA, LIMA, or PEACH WOOD ; *C. brasiliensis*, L., now almost extinct, yields BRAZILETTO-WOOD ; *C. Sappan*, L., SAPPAN, or BUK-KUM-WOOD, and its roots, YELLOW-WOOD or SAPPAN ROOT ; *C. Crista*, L., possibly BRAZIL-WOOD ; *Baphia nitida*, Afzelius, of West Africa, CAM-WOOD or BAR-WOOD ; and *Pterocarpus santalinus*, L. fil., and *Aden-anthera pavonina*, L., both large Indian leguminous trees, yield RED SANDAL or RED SANDERS WOOD. All these Leguminosæ, except Yellow-wood, are used as red dyes.†

FUSTIC or OLD FUSTIC, a fine yellow dye, is the wood of a large West Indian tree of the order Moraceæ, *Maclura tinctoria*, Nutt., whilst the smaller branches of the Anacardiaceous VENETIAN SUMACH, *Rhus Cotinus*, L., are known as ZANTE FUSTIC or YOUNG FUSTIC.

MADDER, the root of *Rubia tinctorum*, L., a British plant, still considerably grown in France, and its Indian ally, MUNJIT, *R. cordifolia*, L., have been already referred to.

CHAY ROOT, *Oldenlandia umbellata*, L., a native of Java and Coromandel, and the INDIAN MULBERRY root, *Morinda citrifolia*, L., and other species possibly, also belonging to the *Rubiaceæ*, are used in India as red dyes, the latter, especially, for red cotton turbans.

TURMERIC, the rhizome of *Curcuma longa*, L., has

* 'Pharmacographia,' p. 186. Bentley and Trimen, ii, p. 86.
† Archer, ' Popular Economic Botany,' pp. 203-207.

been already referred to (p. 63, *supra*) as employed in curry-powder, and no longer used as a dye.

SAFFLOWER, the dried inflorescences of the Composite *Carthamus tinctorius*, L., cultivated in China, India and Southern Europe, is largely consumed in the manufacture of rouge.

SAFFRON, the stigmas of *Crocus sativus*, L., already mentioned, not yielding a permanent yellow, is now but little used as a dye.

YELLOW, or PERSIAN BERRIES, the unripe fruits of the South European *Rhamnus infectorius*, L., are imported mainly from Smyrna, to the extent of about 450 tons annually. The inferior French ones are known as AVIGNON BERRIES.

ORCHELLA WEED consists of several species of the lichen *Roccella ; R. tinctoria*, DC., collected in South America, Morocco, the Canary, Cape Verd, Azores, Madeira and Sardinian islands, *R. fuciformis*, Ach., collected in Madagascar, Angola, Madeira, and South America, and other species. The red or blue liquid ammoniacal extracts are known as ORCHIL or ARCHIL, the dry powder as CUDBEAR, and the alkaline solid preparation, made chiefly in Holland, as LITMUS. The latter is well known as a test for acids and alkalies.*

ARNATTO, the red pulp which covers the seeds of the South American *Bixa Orellana*, L., is used as an orange or yellow dye for silks, and for staining Dutch cheese.

ALKANET ROOT, *Ankanna tinctoria*, Tausch., cultivated chiefly in Southern Europe, and imported on

---

* Bentley and Trimen, iv, pl. 301.

a small scale, is a Boraginaceous plant, used as a red stain for furniture in imitation of rosewood, and, as is also DRAGON'S BLOOD, which was referred to among the resins (p. 119, *supra*), for giving a crimson colour to oils.

QUERCITRON, the bark of the North American *Quercus tinctoria*, L., though useful for tanning, is, with us, mainly employed as a yellow dye.

INDIGO, though obtained by chemical synthesis, is likely for some time to be mainly produced from the Leguminous *Indigofera tinctoria*, L., *I. Anil*, L., *I. cærulea*, Roxb., natives of India, as a precipitate obtained by soaking the plant in water. We imported 4,774 tons of this dye in 1882, as against 3,524 tons in 1851. From it laundress's 'blue' is prepared.

GAMBOGE, though a valuable water-colour, and used also in colouring lacquer varnish, is not a true dye. The best Pipe Gamboge, from Siam, is the gum-resin of *Garcinia Hanburii*, Hook. fil. ; that of Ceylon, of *G. Morella*, Desr. ; and that of Southern India, of other species.

## PART VI.—FIBRES AND PAPER-MATERIALS.

As in oils, gums, resins, rubbers, and tanning materials, so in fibres, the rapidly increasing population of the civilized world and the extended needs of our advancing refinement of life would alone account for the necessity of a search for new material to be carried further afield. Textile fabrics and cordage have for ages created a constant demand upon the supply of fibrous material in the Vegetable Kingdom ; but to these two other channels for the consumption of that material have been added in modern times. There is a constant and considerable demand for certain stalks and fibres of vegetable origin for the manufacture of various kinds of brooms and brushes, and—vastly more important—there is now an enormous demand for fibrous material, either fresh, or in the form of rags, for the manufacture of paper. As many fibres are thus used both for textiles and for paper-making, it will be inconvenient to enumerate them entirely separately. It may be premised that vegetable fibres fall practically under three categories as to their origin. They are either hairs from the seeds, consisting of unaltered cellulose, or they are the ligni-fied bast-fibres or inner bark mostly of Dicotyledons, or, lastly, they may be the entire fibro-vascular bundles of Monocotyledons. With the exception of cotton, derived from the seeds of various species of *Gossypium*, the first class are unimportant, having no staple, and being, accordingly, only used for stuffing cushions, etc., and that, as a rule, only locally.

In volume ix. of the ' Encyclopædia Britannica ' (1879) Mr. James Paton gives a very complete table, compiled from Dr. H. Müller's Pflanzenfaser in Hofmann's 'Bericht uber die Entwickelung der chemischen Industrie.' This enumerates the source, locality, and principal uses of all vegetable fibres hitherto employed for textiles and cordage, etc., to any considerable extent. It includes (*op. cit.*, p. 132) 20 fibres of the first of the above-mentioned three classes, 43 of the second, and 23 of the third ; many of these, however, are only used locally, or have not, at all events, ever been articles of English home commerce. Nevertheless, so rapid has been the advance in this industry that several fibres not included in this list are now in use, and considerable influence in the introduction of wood-pulp as a paper-material and of various new fibres must fairly be ascribed to such exhibitions as the Edinburgh Forestry Exhibition of 1884, and the Colonial and Indian Exhibition of 1886.

In 1837 Sunn-hemp, China-grass, and Rhea and Pineapple fibre were not in use in this country ; whilst the Jute industry at Dundee was on a somewhat tentative scale. In 1851 the fibres just mentioned had been recently introduced, and those of *Marsdenia tenacissima*, Wight and Arn., the ' Jetee ' or ' Rajmahal bow-string hemp,' of *Calotropis gigantea*, Br., and *C. procera*, R. Br., the ' Yercum ' or ' Mudar,' of *Sanseviera zeylanica*, Willd., the " Moorghae ' or ' Marool,' and of *Hibiscus*, were recommended.

Again, in 1851 Paper was made entirely from rags or straw, whilst in 1875 we imported, as against 52,500 tons of rags and pulp, 141,900 tons of Alpha, or so-called Esparto Grass. This latter amount rose the following year to 150,000 tons, and by 1880 to 200,000. In 1877, among materials employed in this manufacture, Professor Archer enumerates* rags, straw, alpha, gunny-bags (jute), and experimentally *Chamærops humilis*, L., *Agave*, *Musa textilis* (Manila hemp), *Phormium tenax*, Forst. (New Zealand flax), *Broussonetia*,

* Bevan's ' Manufacturing Industries.'

Bamboo, Hop-bines, Peat, Bracken and Sawdust. Wood was then scarcely in use here, though employed in Sweden and Belgium.

For a detailed history of paper-materials, with especial reference to wood-pulp, reference may be made to ' Forestry and Forest-products ' (Edinburgh, 1884). It shows that our Patent Office teems with specifications for making paper from all kinds of material, few of which have succeeded. In 1852 Mr. Thomas Routledge introduced the Spanish fibre Alfa, but its general adoption arose from the threatened scarcity of rags during the cotton famine of the American War, which followed immediately upon the repeal of the paper-duty. This fibre cannot apparently be delivered here for less than £4 7s. per ton, so that since 1880, and especially since 1884, the year of the Edinburgh Exhibition, it is being almost completely superseded by wood. Almost any wood can be used; but Aspen and other Poplars, Alder, Spruce and Pine seem the most suitable, either from quality or quantity available. There may be a future demand for many cheap fibres, or for more expensive ones, if they can first be used for textiles, for ' bagasse ' or refuse cane, or for sugar-cane refuse ; but wood seems likely to take the lead as a paper-material for many years to come. Meanwhile the applications of paper, long so varied in Japan, are greatly extending with us. At Cleobury-Mortimer, Shropshire, coach-panels and tram-wheels and black ' patent leather ' is made from it, whilst at Willesden it is water-proofed by means of cupro-ammonium for roofing-purposes.*

* ' Paper Trade Review,' May 23rd, 1884.

For cordage, at the present time, in addition to the longer-known fibres, hemp, jute, Manila hemp, Sunn hemp, and New Zealand flax, the four following fibres are commonly employed.

MEXICAN GRASS, from *Agave americana*, L., and other species, shipped from Tehuantepec, of which 19,000 cwt. have been imported.

SILK GRASS, *Agave vivipara*, L.

PITA, *Bromelia Karatas*, L., of Mexico, and GRASS or SISAL HEMP, *Agave sisalana*, Engelm., from Yucatan, which is now grown in Jamaica, and in London is worth from £5 to £10 less than Manila hemp.

The following alphabetical list* refers only to the remaining more important fibres used for textiles, cordage, or paper:

*Adansonia digitata*, L., BAOBAB, from Angola, in use for paper, £9 to £15 a ton.

*Agave americana*, L., PITA, mixed with Manila hemp for cordage, matting, and the best papers, £7 to £10 a ton less than Manila hemp.

*Ananas sativus*, Mill., PINEAPPLE, from China, Malacca, Java, Philippines, Mauritius, etc., for good strong paper.

*Areca catechu*, L., spathe suggested for paper.

*Arenga saccharifera*, Labill., GOMUTI PALM. Ramenta, or VEGETABLE BRISTLES, suggested as a covering for telegraph cables.

*Attalea funifera*, Mart., BAHIA PIASSAVA. Ramenta worth £15 to £25 a ton for brush-making or cordage, and we import the bulk of the 5,000 tons annually exported.†

* Mainly from Spon's 'Encyclopædia of Industrial Arts.'
† 'Pharm. Journ.,' ix (1850), p. 431.

*Bambusa arundinacea,* L., and *B. vulgaris,* Wendl., yielding 5 to 10 tons of dry young shoots per acre ; recommended by Mr. Routledge for paper.

*Bœhmeria nivea,* Hook. and Arn., *B. Puya,* Hook., of Nepal, and perhaps other species, CHINA GRASS, ORTIE DE CHINE, RAMIE, RHEA, capable of being grown in the climate of Cornwall, a substitute for cotton, the best flax, wool, silk, Pita, Manila hemp, or Phormium. Too dear for paper.

*Broussonetia papyrifera,* Vent., and *B. Kœmpferi,* Sieb., PAPER MULBERRY, used in Japan for paper, and a little of it here in the best plate-paper ; but costing £30 a ton.*

*Calotropis gigantea,* Br., YERCUM or MUDAR ; recommended for paper.

*Cannabis sativa,* L., HEMP, a native of Northern India and temperate Asia, cultivated for centuries in warm and temperate regions, is mainly used for cordage. It is imported mostly from Russia, but also from India, Italy, etc.

*Caryota urens,* L., KITTOOL or INDIAN GUT. The ramenta have been imported since 1860 from Colombo, at 3½d. to 10d. a pound, for brushes.

*Cocos nucifera,* L., the COCOA-NUT, from the mesocarp, or fibrous husk, of the fruit of which is obtained the valuable fibre known as COIR, used for cordage in India, and imported in enormous quantities for matting from Ceylon, Bombay, etc.

*Corchorus capsularis,* L., and *C. olitorius,* L., JUTE, was manufactured tentatively in the last century at Abingdon, and between 1825 and 1840 at Dundee. From the latter date it has formed the staple industry

* R. H. Collyer, ' Rheea Fibre,' Calcutta, 1880, 8vo.

of that town and of Arbroath, which are now, however, seriously rivalled by Dunkirk and Calcutta. In 1851 we imported 21,000 tons ; in 1866, 81,300 tons ; in 1873, 231,245 tons; declining in 1875 to 170,830 tons, but rising again to 232,032 tons, valued at from £11 to £22 a ton, of which, consequently, only the waste is available for paper.*

*Crotalaria juncea*, L., the SUNN HEMP of India, a Leguminous annual, extensively cultivated throughout the country.

*Daphne papyracea*, Wall., the NIPÂL PAPER-SHRUB, suitable for plate-paper.

*Gossypium herbaceum*, L., *G. arboreum*, L., and *G. barbadense*, L., Malvaceous plants, have on the outer surface or testa of their seeds the unicellular hairs, composed almost entirely of cellulose ($C_6H_{10}O_5$), known as COTTON. *G. herbaceum*, L., was a native of India. It has very hairy leaves, divided into three, five, or rarely seven lobes, and flowers yellow with a purple centre, or rarely wholly yellow, white or purple. *G. arboreum*, L., a native of Tropical Africa, has thick, glossy, deeply five or seven lobed leaves, and purple flowers, often with a yellow centre. *G. barbadense*, L., apparently the ancestral type of all the cottons cultivated in America at the time of its discovery, has almost smooth, three to five lobed leaves with heart-shaped base, and yellow flowers with a crimson spot. All three species are now cultivated in India, Egypt, and America, and there are numerous hybrids and varieties. Sanskrit records carry back the use of cotton to 800 B.C., and it may have been

* 'Pharm. Journ.,' ix (1850), p. 545.

equally long known in Egypt, Nineveh, and Peru. In 1886 we retained for consumption and manufacture 671 026 tons of cotton, or more than 40 lb. per head of the population, and we exported cotton yarn to the value of nearly 11½ millions sterling, and cotton manufactures valued at over 57 millions sterling. The refuse seeds yield a good oil-cake for cattle.*

*Hibiscus esculentus*, L., and *H. cannabinus*, L.; recommended.

*Leopoldinia Piassaba*, Wallace. Pará Piassaba, of which the ramenta are used for brushes, and fetch £25 to £45 per ton.

*Linum usitatissimum*, L., FLAX, already mentioned (pp. 81 and 133, *supra*), for the valuable oil obtained from its seeds, was cultivated for fibre in prehistoric times, as evidenced by remains in the Swiss lake-dwellings, and has been grown for ages in Egypt. It will grow equally in very warm and in far colder climates; and though we import more than half our supply from Russia, flax is one of the staple crops of Ireland. We import nearly 100,000 tons of Flax and Tow, or CODILLA OF FLAX, the rough or broken fibres or combings, while we export linen yarn and manufactures to the value of over six millions sterling. Being a true bast or lignified fibre, flax is far stronger than cotton, and linen rags are still one of our chief paper-materials.†

*Macrochloa tenacissima*, Kunth. (= *Stipa*), the entire

* J. Forbes Royle, ' On the Culture and Commerce of Cotton,' 1851 ; Archer, ' Popular Economic Botany,' pp. 170-181 ; ' Cata- ogue of Indian Exhibits : Colonial and Indian Exhibition,' pp. 124-126.
† ' Pharmacographia,' p. 91 ; Archer, ' Popular Economic Botany,' p. 148.

source of Alfa or Esparto, which is not derived from *Lygeum spartum*, L. It was introduced in 1856 by Mr. Thomas Routledge, and sells at £6 to £10 per ton.*

*Musa textilis*, Nees, Manila hemp, of which we imported 407,000 cwt. in 1880, mainly for cordage, and worth from £20 to £70 per ton.

*Phormium tenax*, Forst., introduced from New Zealand in 1822, and unsuccessfully planted in Ireland, but now grown successfully in the Orkney Islands. The demand exceeds the supply, and though there are several varieties or qualities of the fibre, better rope can be made from it even than from *Musa textilis*, and it sells at from £17 to £22 per ton, the refuse only being used for paper.†

*Sida rhombifolia*, L. (*retusa*), Queensland Hemp, which grows like a weed, and is suitable for paper, but costs £30 per ton.

*Tilia europœa*, L., the LINDEN or LIME, a British tree, belonging to the same Natural Order as the Jute, has an abundant, tough and flexible inner bark or BAST; largely imported from Russia as RUSSIA MATTING.

*Tillandsia usneoides*, L., Spanish Moss or Vegetable Horsehair, shipped from New Orleans, mainly for stuffing.

Mr. Christy has recently recommended‡ the fibre of *Musa Ensete*, Gmel., of Abyssinia, which was rediscovered by Mr. Plowden in 1853. In that year also the source of the interesting Rice-paper of China was ascertained as *Fatsia papyrifera*, Decaisne.

* L. Bastide, ' L'Alfa,' Oran, 1877, 8vo.
† W. Gibbs, ' New Zealand Flax,' London, 1865, 8vo.
‡ ' New Commercial Plants,' No. 9, p. 25.

At the time of the Crimean War various substitutes were introduced for the Russian Linden Bast, so largely used by gardeners and florists. Among these were CUBA BAST, from *Paritium elatum*, G. Don., and RAFFIA, the epidermis of the leaves of *Raphia Ruffia*, Martius, a Palm of Madagascar, reaching us, *viâ* Mauritius, in strips, ¼ to ⅜ in. wide, and also from *R. tædigera*, Martius, of Brazil, worth £40 or £50 a ton. The former is now disused.

Among materials used in the manufacture of brushes, besides the 'ramenta,' or fibres from the petioles of various Palms, to which allusion has already been made, we may mention two: BRUSH-GRASS or 'Chiendent,' the wiry rhizomes of *Chrysopogon Gryllus*, Trin., a native of the South of Europe; and BROOM-CORN, or 'Brush,' the dried rachides of the panicles of the species of *Sorghum* cultivated at Schenectady, Montgomery, and elsewhere in the United States, from 1843, especially by the Shaker communities.*

## PART VII.—TIMBER AND OTHER WOODS.

THE greatly increased use of iron has by no means diminished the demand for wood for purposes of domestic architecture, though it has undoubtedly done so in the case of that required for ship-building. At the same time, the natural growth of population and increase of houses and ordinary expansion of the trades employing wood have caused a greatly in-

* P. L. Simmonds, 'Common Products of the Vegetable Kingdom' (1854), and Spon's 'Encyc. Indust. Arts,' p. 543.

creased demand, especially for the soft woods of the Baltic and of Canada. The increase of mining has caused the home supply of pit-props to fall far short of the demand ; whilst whole tracts of forest in Sweden are now annually cleared by the require-ments of the wooden match-manufacture alone.

Recklessness in the past had in 1875 reduced the valuable forests of Box-wood in Mingrelia to the verge of extinction, and the costliness of the wood has caused, in spite of the competition of other methods of engraving, an anxious search for some efficient substitute. Dividing woods into those re-quired for this purpose, those used for walking-sticks, etc., those used for ornamental purposes, veneering and furniture, and those mainly employed as timber in works of construction, we may first transcribe the list of about 25 proposed substitutes for Box, given in a valuable paper contributed to the ' Journal of the Society of Arts,' in 1886, by Mr. John R. Jackson, the able Curator of the Kew Museums.

### PITTOSPOREÆ.

*Pittosporum undulatum*, Vent., Australia, where it is know as ' BOX.' Introduced experimentally in 1862, but inferior.

*P. bicolor*, Hook., New South Wales and Tasmania : superior to the former.

*Bursaria spinosa*, Cav. : equal to inferior Box, but Bblunts tools ; the OX-TREE of Tasmania.

### MELIACEÆ.

*Swietenia mahagoni*, L. : only suited for coarse work.

### ILICINEÆ.

*Ilex Opaca*, L., North America : useful for turning, but inferior to *Cornus florida*, L., for engraving.

### CELASTRINEÆ.

*Elæodendron australe*, Vent., of Australia : used there for coarse work.

*Euonymus Sieboldianus*, Blume, PAI'CHA, of China : introduced in 1878, very useful, though inferior to Box, Pear, and Hawthorn.

### SAPINDACEÆ.

*Acer saccharinum*, L. : much used for cabinet-making and furniture, but not found suitable by engravers.

### LEGUMINOSÆ.

*Brya ebenus*, A. DC., GREEN EBONY of Jamaica, or COCUS WOOD : used for flutes, etc. ; equal to bad Box.

### ROSACEÆ.

*Pyrus communis*, L. : used for carving and calico-printers' blocks, but inferior to Hawthorn.

*Amelanchier Sieboldianus*, L., SERVICE TREE of America : recommended for trial.

*Cratægus Oxyacantha*, L., HAWTHORN : next best to box of any known wood, but probably difficult to obtain of any large size.

### MYRTACEÆ.

*Eugenia procera*, Poir., of the West Indies : suited for coarse engraving.

### CORNACEÆ.

*Cornus florida*, L., AMERICAN DOGWOOD or AMERICAN BOX : abundant, but only used for coarser engraving.

## ERICACEÆ.

*Rhododendron maximum*, L., *R. californicum*, Hook., and *Kalmia latifolia*, L., have all been strongly recommended in America.

## EPACRIDACEÆ.

*Monotoca elliptica*, R. Br., of Australia : found unsuitable.

## EBENACEÆ.

*Diospyros texana* : small, but nearly equal to the best box.

*D. virginiana*, L. : used for shuttles.

*D. ebenum*, Kœnig : nearly equal to Box.

## APOCYNACEÆ.

*Hunteria zeylanica*, Gaertn. : used in Ceylon, and useful, if sufficiently abundant.

## BIGNONIACEÆ.

*Tecoma pentaphylla*, DC., of Brazil and the West Indies : the most 'likely successor to Box,' called WEST INDIAN BOX-TREE.

## BUXINEÆ.

*Buxus Macowanii*, Hook. fil., the CAPE BOX, is now considerably used, and closely resembles *B. sempervirens.* It is not, however, plentiful in South Africa.

## CORYLACEÆ.

*Carpinus Betulus*, L., the HORNBEAM, and *Ostrya virginica*, Willd., have been recommended.

Our knowledge of the botanical sources of the bamboos, canes, and sticks used at the present day for walking-sticks, umbrella-handles, etc., is most imperfect. Such as it is, it has been ably summarized by Mr. J. R. Jackson.* Only twenty years ago our imports, to supplement home-grown oak, ash, blackthorn and holly, consisted almost exclusively of bamboos and of Partridge and Tonquin canes. We now import more than fifty different species, from Singapore, China, the West Indies, Algeria, and the Continent of Europe, twenty-nine million sticks coming annually from the East, three million from Europe, and two million from Algiers, the total import being valued at £300,000. Among exceptional structures used for these purposes are the JERSEY CABBAGE, a tall variety of *Brassica oleracea*, L., produced by stripping off the lower leaves, light but weak; the so-called 'THISTLE,' the stems of the MULLEIN (*Verbascum Thapsus*, L.), a British plant; the TEAZLE, the fasciated and consequently twisted stalks of the FULLER'S TEAZLE (*Dipsacus fullonum*, Mill.), imported in large numbers from the South of France, but having no strength; and the triangular midribs of the leaves of the DATE-PALM (*Phœnix dactylifera*, L.), imported from Algiers. Those most commonly cut out of the solid log are OAK (*Quercus Robur*, L.) home-grown; CEYLON or MACASSAR EBONY (*Diospyros Ebenum*, Kœnig); and the PALMYRA PALM (*Borassus flabelliformis*, L.), of India. Those most commonly grown in England are ASP, or APSE (*Populus tremula*, L.), valued for ligthness; ASH

* 'Journ. Soc. Arts,' 1888.

12

(*Fraxinus excelsior*, L.); BIRCH (*Betula alba*, L.);
BLACKTHORN (*Prunus spinosa*, L.), also imported
from Ireland; CRAB (*Pyrus Malus*, L.); FURZE
(*Ulex europæus*, L.); HOLLY (*Ilex Aquifolium*, L.),
much used for driving-whips; HORNBEAM (*Carpinus
Betulus*, L.); MAPLE (*Acer campestre*, L.); MOUN-
TAIN ASH (*Pyrus Aucuparia*, L.); OAK and WHITE-
THORN (*Cratægus Oxyacantha*, L.). Among those
imported from the Continent are CHERRY (*Prunus
Cerasus*, L.), from Austria and Hungary, especially
for pipe-stems; CHESTNUT (*Castanea sativa*, Scp.),
from France; CORK OAK (*Quercus Suber*, L.), from
Spain; DOGWOOD (*Cornus sanguinea*, L.; TEAZLE,
GUELDER ROSE, or BALKAN ROSE (*Viburnum
Opulus*, L.), from the Balkans; HAZEL (*Corylus
Avellana*, L.); and MEDLAR (*Pyrus germanica*, L.),
from France. From Algeria we get the CAROB, or
CARONBIER (*Ceratonia Siliqua*, L.); the CORK OAK,
DATE PALM; EUCALYPTUS (*Eucalyptus Globulus*, Lab.),
specially cultivated for this purpose; MYRTLE, pos-
sibly *Myrtus communis*, L.; NANA CANES (*Arundo
Donax*, L.); OLIVE (*Olea europæa*, L.); ORANGE
(*Citrus*, spp.); BLACK ORANGE, a curious trade name
for the Broom (*Cytisus scoparius*, Link.); and POME-
GRANATE (*Punica Granatum*, L.); besides the BAY-
TREE, or LAURIER THYN, some species of *Eugenia*.
From the East Indies we get numerous RATTAN
(*Calamus*, spp.) and TONQUIN (*Arundinaria*, spp.)
CANES, BAMBOOS, species of *Bambusa* and *Arundi-
naria*; from China, the so-called CAROLINA REED,
also an *Arundinaria* from the same country, the most
valuable MALACCA CANE (*Calamus scipionum*, Lour.,

the PALMYRA PALM, the undetermined PARTRIDGE CANE of China, the PENANG LAWYER (*Licuala acutifida*, Mart.), the RAJAH CANE (*Eugeissonia minor*, Beccari) of Borneo, the WHANGEE BAMBOO (*Phyllostachys nigra*, Sieb. and Zucc.) of Japan, and EBONY from Ceylon. BOXWOOD, obtained from Persia, is used for sticks as well as for engraving. From Australia we have the LOYA CANES (*Calamus australis* Mart.), the MIDGEN (*Kentia monostachya*, F. von Muell.), a palm, and the scented MYALL (*Acacia homalophylla*, A. Cunn.), used for pipes. From the West Indies, the BEEF-WOOD of Cuba, probably *Ardisia coccinea;* the BRIAR (*Zanthoxylum Clava-Herculis*, L.), the little-used CEDAR (*Juniperus virginiana*, L.), the COFFEE (*Coffea arabica*, L.), the FLOWERED EBONY, or COCUS (*Brya Ebenus*, A.DC.), the PARTRIDGE-WOOD (*Andira inermis*, Kunth.), the PIMENTO of Jamaica (*Pimenta officinalis*, Lindl.), and various doubtful woods, such as BLACK FORK, and GRU-GRU, a palm. The valued LANCE-WOOD, used for shafts, fishing-rods, bows, etc., at once strong and elastic, is *Duguetia quitarensis*, Benth., of Guiana ; the SNAKE-WOOD, also known as LETTER-WOOD or LEOPARD-WOOD, is *Brosimum Aubletii*, Pœpp.; and the BRAZILIAN OAK, or CEYLON VINE, imported from Bahia, is quite undetermined.

As with canes, so with ornamental and even long-known timber-woods, many kinds are very imperfectly known to botanists, being sent down to the coast without flowers, leaves, or even bark to aid in their determination. The most important woods used for furniture, veneering, and other ornamental purposes are :

12—2

OAK (*Quercus Robur*, L.), largely grown in England, and also imported, that from Riga being the best, owing to the 'figuring' produced by its medullary rays. It is used in veneering.* The WHITE OAK (*Q. alba*, L.) of North America is less valuable.

PITCH PINE (*Pinus australis*, Michx.), imported from Darien, Savannah, etc., is, next to oak, the wood that enters most largely into church, school, and house fittings, the better specimens being used in cabinet-work.

MAHOGANY (*Swietenia Mahagoni*, L.), first imported from Honduras about 1725, but now mainly obtained from Cuba, under the name of SPANISH MAHOGANY, has long been a favourite furniture wood, and is much used in veneers. It is also imported from Tabasco and other parts of Mexico.

BEECH (*Fagus sylvatica*, L.), besides being considerably grown, especially for seats of chairs, is also largely imported, mainly from Hamburg and, in a manufactured state, from Vienna, where it is used in bentwood furniture. It can be readily stained.

BIRCH (*Betula alba*, L.) is more used on the Continent than with us, but increasing quantities are imported from the Baltic to our manufacturing districts. The AMERICAN or QUEBEC BIRCH (*B. lenta*, L.) is more used as a furniture wood, and when 'figured' or 'curled' is cut into veneers. It is now largely imported in planks.

WALNUT (*Juglans regia*, L.) was much in demand at the beginning of the century for gun-stocks, for which the Italian-grown wood is still largely used.

* Thomas Laslett, 'Timber and Timber Trees,' 1875, p. 96.

It is also imported from the Black Sea and from France, English-grown wood being pale, coarse, and inferior. Burrs or excrescences frequently occur on the trees, which yield mottled wood much valued for veneers for pianos and other furniture. AMERICAN WALNUT (*J. nigra*, L.), a more durable but more monotonous and duller wood, is now more used than *J. regia* for cabinet-work.

ASH (*Fraxinus excelsior*, L.), much used in carriage-building, is now also largely superseded for coach-panels, cabinet-drawers, and furniture by the AMERICAN, QUEBEC, or WHITE ASH (*F. americana*, L.).

BIRD'S-EYE MAPLE (*Acer saccharinum*, Wang.), a spotted variety, is considerably imported as veneers from the New England States. BOG MAPLE is the same wood turned pale blue from being buried in peat.

SATIN-WOOD (*Chloroxylon Swietenia*, DC.), imported from Nassau, in the Bahamas, is one of our most beautiful but scarcest furniture-woods.

CUBA, HONDURAS, or MEXICAN CEDAR (*Cedrela odorata*, L.) is not a coniferous tree. It is used for cigar-boxes and the interior of natural history cabinets, its odour being obnoxious to insects.

PENCIL CEDAR, the similarly-coloured woods of *Juniperus virginiana*, L., and *J. bermudiana*, L. are now only used for lead-pencils.

ELM (*Ulmus campestris*, With., and *U. montana*, Sm.), once used for linen-chests, is now little employed save for coffins. Its burrs are, however, very ornamental.

HOLLY (*Ilex Aquifolium*, L.) is used for inlaying,

especially for ' bands ' and ' strings ' and in Tunbridge ware ; and, since it takes stains readily, it is also used for imitating ebony.

SYCAMORE (*Acer Pseudo-platanus*, L.), known as ' Plane ' in Scotland, is a white wood, considerably employed in turnery, for bread-platters, butter-moulds, etc. ; but the true PLANE (*Platanus occidentalis*, L.) of America, there known as BUTTON-WOOD, LACE-WOOD, or HONEYSUCKLE-WOOD, is prettily figured, and is coming into use for furniture and veneering.

POPLAR, WHITE, YELLOW, or VIRGINIAN POPLAR, WHITEWOOD, or CANARY WHITEWOOD, are commercial names for the wood of the TULIP - TREE (*Liriodendron tulipifera*, L.) of the United States, now coming into extensive use in carriage-building, cabinet-making, and shop-fitting.

SWISS PINE is similarly the trade name for the wood of the SILVER FIR (*Abies pectinata*, DC.) ; the most sonorous of woods, which is imported from the Continent for the bellies of violins and the sounding- boards of pianos.

The Colonial and Indian Exhibition has called attention to various useful woods of this class, such as *Cedrela Toona*, Roxb., the TOON of India, THIT-KADO of Burma, MOULMEIN CEDAR, or INDIAN MAHOGANY of commerce ; PORCUPINE-WOOD, that of the COCOANUT PALM (*Cocos nucifera*, L.) used in inlaying ; INDIAN ROSEWOOD (*Dalbergia latifolia*, Roxb.) ; and especially *Pterocarpus indicus*, Willd., the PADOUK, ANDAMAN REDWOOD, or BURMESE ROSEWOOD, a handsome dark-red wood among Indian timbers ; the MIRABOO (*Afzelia palembanica*), RASSAK

(*Vatica Rassack*), and other species botanically undetermined, such as BILLIAN and COMPASS from Borneo,
TRINCOMALEE WOOD (*Berrya Ammomilla*, Roxb.)
from Ceylon, the STINKWOOD (*Ocotea bullata*, Nees),
and other valuable timbers from Cape Colony; the
SANTA MARIA (*Calophyllum Calaba*, Jacq.) and AN-
GELIN (*Andira inermis*, Kunth) of the West Indies,
and many others. There are still, however, unfortu-
nately, many of these which are not botanically deter-
mined. Such are the valuable AMBOYNA WOOD
from the Malay Archipelago, probably a species of
*Pterocarpus*, also known as KIABOOCA WOOD, the
various species of *Dalbergia*, and other genera from
both East and West, known as ROSEWOODS, and to a
less extent the EBONIES and SANDALWOODS.

*Diospyros Embryopteris*, Pers., of India, and *D. Ebenum*, Kœn., of Ceylon, yield the true EBONY. *D.
quæsita*, Thw., is the CALAMANDER WOOD; but other
species, such as *D. melanoxylon*, Roxb., from India, and
from Tropical Africa, produce similar valuable woods.

*Santalum album* L. is the SANDAL-WOOD of India,
*S. Freycinetianum*, Gaudich. and *S. paniculatum*, Hook.
and Arn. that of the Sandwich Islands, *S. Yasi*,
Seem., that of Fiji, *S. austro-caledonicum*, Vieill.,
of New Caledonia, and *S. spicatum*, A.DC. (=
*Fusanus*) of West Australia.

As to timbers used for construction, we have information in the papers by the veteran technologist, Mr.
P. L. Simmonds, in the 'Nautical Magazine' for 1874
and 1875, and in the late Mr. Thomas Laslett's
masterly work, 'Timber and Timber-trees' (1875).
Mr. W. Stevenson has reprinted from the 'Timber

Trades Journal' (1888) a valuable series of articles on 'The Trees of Commerce;' and the present writer, in addition to the mention of the uses of the species grown in England in 'Familiar Trees' (Cassell and Co., 1885-88), has compiled a list of the chief timbers of the world, arranged geographically, with their scientific and vernacular names, under the title of 'Economic Forestry' ('Transactions of the Scottish Arboricultural Society,' vol. xi., pp. 382-480, 1887). Of these many of the chief have been already mentioned under the two preceding classes, with the exception of the Conifers.

*Pinus sylvestris*, L., under the names of DANTZIC or RIGA FIR, REDWOOD, RED or YELLOW DEAL, is, perhaps, imported in larger quantities than any other wood. That from the White Sea and St. Petersburg, and that shipped from Gefle and Soderhamn in Sweden, is of fine quality ; but the bulk of that imported comes from Dantzic and Riga. The timber of old native Scots Firs in the North of Scotland is beautifully figured, and has been much used for panelling at Balmoral. The uses of this wood are infinite —masts, deck-planks, and beams in ship-building, railway-sleepers, wood-paving, scaffolding, rafters, and flooring, are among the chief.

*P. resinosa*, Aiton, the RED or RESINOUS PINE of commerce, a native of Canada and the United States, is used in our dockyards, but not imported in large quantities.

*P. mitis*, Michaux, the AMERICAN SOFT or YELLOW PINE, or NEW YORK PINE of commerce, largely used in American dockyards, is also sparingly imported into England.

*P. Strobus*, L., the AMERICAN WHITE PINE of commerce, the YELLOW PINE of our dockyards, the WEYMOUTH PINE of botanists, introduced by Lord Weymouth at Longleat in the last century, is more used in America than any other pine, especially in house-building. In England it is imported from Quebec, and used for masts and spars, carriage-building, cabinet-making, and house-building, for which it is unrivalled.

*P. australis*, Michaux, the PITCH PINE, a native of the Southern United States, is the most extensively imported of American timbers. It is used for masts, spars, beams, pit-props, wood-paving, wainscot, and as already mentioned (p. 180 *supra*), furniture, and furnishes most of the pitch and turpentine of the world.

*P. Pinaster*, Solander (*P. maritima*, Poiret), the CLUSTER PINE of Southern France, is imported to Cardiff for mining timber.

*Picea excelsa*, Link., the NORWAY SPRUCE or WHITEWOOD, is extensively imported, the largest timber coming from Riga, Memel, and Dantzic, the best from St. Petersburg and Archangel. Whole trees are used for scaffold-poles, ladders, and small masts; the whiteness of the wood causes it to be used for kitchen tables and flooring; its cheapness, for packing-cases and firewood; and it is also used for paper-pulp.

*P. alba*, Link., and *P. nigra,* Link., and *Tsuga canadensis*, Carrière, the WHITE, BLACK, and HEMLOCK SPRUCES of North America, are largely imported from New Brunswick and the rest of Canada, especially for carpenters' work.

*Larix europæa*, DC., the LARCH, is exceptionally

durable, and is used on the Continent for water-pipes, casks, beams, and cabinet-work ; and in England for hop-poles, fencing, pit-props, tramway-sleepers, and wheelwrights' work. Affording a rapid return] to capital, even on poor mountain-land, no tree is now so extensively planted in Britain.

*Pseudotsuga Douglasii,* Carrière, the OREGON PINE, or DOUGLAS or NOOTKA FIR, abundant in North-west America, furnishes fine, straight, and durable timber, valuable for spars, but, from occasional want of cohesion between the annual layers, not so good for topmasts as Dantzic or Kauri Pine. The flagstaff at Kew, of this wood, is 159 feet long. Though more rapid in its growth, and probably, therefore, less durable in Scotland than in Oregon, this species seems in many respects more valuable than Larch.

*Dammara australis,* Lambert, the KAURI or COWDIE PINE of New Zealand, largely used there and in Australia, is for size, lightness, elasticity, strength, and durability unrivalled, whether for masts or decks. It also yields abundance of the valuable Kauri gum or resin.

In 1851 the eight timbers classed A1 for ship-building at Lloyd's were OAK (*Quercus Robur,* L.), LIVE OAK (*Q. virens,* Ait.), AFRICAN OAK (*Swietenia senegalensis* Desv.), TEAK, SAL, GREENHEART, MORA, and IRON-BARK, of which GREENHEART (*Nectandra Rodiæi,* Schomb.) and MORA (*Mora excelsa,* Benth.), had then been but recently introduced from British Guiana ; whilst IRON-BARK (*Eucalyptus resinifera,* Sm.) was only added to the list during the Exhibition.

*Quercus virens,* Aiton, the AMERICAN LIVE OAK, an evergreen, growing near the coast, especially in Florida, yields small and crooked timbers, dark-coloured, hard, tough, heavy, with a twisted grain, and stronger than any other oak. It is largely used in American dockyards, and for mallets and cogs.

*Swietenia senegalensis,* Desv., AFRICAN OAK, TEAK or MAHOGANY, a dark-red, very hard, strong, fine-grained wood, is imported from Sierra Leone, and used for keels, beams, etc.

*Tectona grandis,* L. fil., the TEAK, a native of Southern India, Burmah, and Siam, thriving best with an annual rainfall of from 50 to 120 inches, is a lofty tree, frequently 100 feet high, and six to ten feet in circumference. The wood is moderately hard and strong, almost indestructible under ordinary circumstances, and contains a resinous oil which resists insects, water, and rust, for which reason it is the best wood known for backing armour-plates in ironclads. The Teak forests of Burmah extend from near the exporting ports of Rangoon and Moulmein to the Shan States and the Chinese frontier. The wood is largely used in India, and about 42,000 loads annually imported into Great Britain.*

*Shorea robusta,* Gaertn., the SAL, is the most-used timber of Northern India—brown, hard, and durable.

*Nectandra Rodiœi,* Schomb. the GREENHEART, dark green or black, remarkably strong and durable, resisting the white ant and the boring *Limnoria*

* See W. T. Oldrieve and John C. Kemp in 'Forestry and Forest Products,' pp. 321-406.

*terebrans,* is largely used in our dockyards for keels, beams, etc., as also is

*Mora excelsa,* Benth., MORA, which reaches a height of 150 feet, yielding timber twelve to twenty inches square and thirty-five feet long. Its wood is a handsome chestnut-brown, beautifully figured.

*Eucalyptus resinifera,* Sm., and no doubt other species of this magnificent Australian genus, have been imported as IRONWOOD, a deep-red, straight-grained, very hard and heavy wood, useful for the same purposes as the two last-mentioned species.

Among numerous proposed teak substitutes we can only mention the TEWART (*Eucalyptus goniocalyx,* F. von Muell.) of West Australia ; the CHOW or MENKABANG PENANG (*Casuarina equisetifolia,* L. fil.) and MIRABOO (*Afzelia palembanica,* Baker) of Borneo ; the MERANTI (*Hopea Meranti*) and BINTANGORE or POON (*Calophyllum Inophyllum,* L.) of the Straits Settlements ; the PYNKADO (*Xylia dolabriformis,* Benth.), THINGAN (*Hopea odorata,* Roxb.), NAN-TA-ROOP (*Altingia excelsa,* Noronha), and PADOUK (*Pterocarpus indicus,* Willd.) of Burmah ; the JARÚL (*Lagerstrœmia Flosreginœ,* Retz.), CHAMPA (*Michelia Champaca,* L.), and SISSOO (*Dalbergia Sissoo,* Roxb.) of India ; and the ANGÉLIQUE or ANGELIN, (*Andira inermis,* Kunth.) of Guiana.

Since 1851, though for furniture purposes American WALNUT, ASH, MAPLE, TULIP-TREE or WHITEWOOD, and the beautiful REDWOOD of California (*Sequoia sempervirens,* Endl.), have been largely used, attention has been mainly attracted for ship-building purposes by the SABICU (*Lysiloma Sabicu,* Benth.) of

Cuba, of which the Exhibition stairs were made ; by the HUON PINE (*Dacrydium Franklinii*, Hook. fil.), and various other fine woods of Tasmania ; by the TOTARA (*Podocarpus Totara*, A. Cunn.), KAURI PINE, and other excellent woods with astringent barks in New Zealand ; and, more recently, by the enormous and abundant JARRAH of West Australia (*Eucalyptus marginata*, Sm.), which rivals Spanish Mahogany in beauty, and by the other members of that remarkable genus.

Our limits do not permit reference to the many beautiful Conifers and shrubs with which our gardens have been enriched.

---

## PART VIII.—AGRICULTURAL PLANTS.

IN addition to the cereals, pulses, and other plants used for human food, and the locally-cultivated medicinal plants, the chief agricultural crops are those species required as food for cattle. Of these, Oats, Rye, Peas, Field-Beans, Maize, and Turnips have been already mentioned. (See Part I.) It remains to say a few words as to Swedes, Mangold-Wurzel, Kohl-Rabi, Grasses, Clovers, and other fodder-plants.

The SWEDE (*Brassica campestris*, L., var. *Napobrassica*, DC.), or Swedish Turnip, is a most valuable food for sheep, yielding far heavier crops than the common Turnip, the introduction of which, in the seventeenth century, effected a revolution in our agriculture. Swedes flourish in the moist summers of the North of Britain, but suffer much from the attacks of the Turnip Beetle.

MANGOLD-WURZEL (*Beta vulgaris*, L., var. *Cycla*), of nearly equal feeding value, yielding heavier crops, and not liable to the attacks of the Turnip Beetle, is adapted for hot and dry conditions where the Turnip will not flourish, as on the light soils of our southern counties.

KOHL-RABI is a variety of the Cabbage (*Brassica oleracea*, L.), in which the base of the stem is enlarged into a green turnip-like mass bearing the sheathing bases of the leaves. Sheep and cattle are fond of it.

There is in Britain a very large acreage of natural meadow-land, and careful studies have been made by Messrs. Lawes and Gilbert, Buckman, Sutton, and others as to the most valuable hay-yielding grasses which this meadow-land yields, either with or without irrigation. These studies have furnished suggestions as to the most suitable species to sow for the formation of permanent pasture, or for alternate husbandry. Among these are the following grasses: RYE-GRASS (*Lolium perenne*, L.), and its variety, ITALIAN RYE-GRASS (*L. perenne*, var. *italicum*), especially for hay, COCK'S-FOOT GRASS (*Dactylis glomerata*, L.), TIMOTHY or CAT'S-TAIL GRASS (*Phleum pratense*, L.), especially for heavy or peaty land, *Anthoxanthum odoratum*, L., especially for hay, *Poa pratensis*, L., for dry, sandy soils, *P. nemoralis*, L., var. *sempervirens*, and *P. trivialis*, L., for permanent pasture ; and the FESCUES, *Festuca duriuscula*, L., *F. elatior*, L., and *F. pratensis*, Hudson, for permanent pasture, and *F. rubra*, L., on sandy soils. Besides these the following Leguminous plants are recommended for sowing: SAINFOIN (*Onobrychis sativa*, Lam.), especially on dry calcareous soil, WHITE or DUTCH CLOVER (*Trifolium repens*, L.), RED CLOVER (*T. pratense*, L.), and its perennial variety, ALSIKE CLOVER (*T. hybridum*, L.), introduced during the present century, and YELLOW CLOVER, BLACK MEDICK or NONSUCH (*Medicago lupulina*, L.). BURNET (*Poterium Sanguisorba*, L.), a Rosaceous plant, on calcareous soils, MILFOIL (*Achillea Millefolium*, L.), a Composite, PARSLEY (*Petroselinum sativum*, Hoffm.), an Umbellifer, and the RIBWORTS (*Plantago lanceolata*, L., and on calcareous soil *P. media*, L.), are recommended in addition for sheep pastures, and the CHICORY (*Cichorium Intybus*, L.) for cattle. CRIMSON CLOVER or TRIFOLIUM (*Trifolium incarnatum*, L.) is a valuable forage plant, and makes good hay; and the large RED CLOVER (*T. pratense*, L., var. *sativum*), either alone or with Italian Rye-grass, is one of the chief fodders used for stall-fed cattle. The VETCH (*Vicia sativa*, L.) has long been a favourite crop on all soils ; but the LUCERNE (*Medicago sativa*, L.), requiring a light soil and dry subsoil, is less generally used. The young shoots of the GORSE, WHIN, or FURZE (*Ulex europæus*, L.), a spinous hardy plant, flourishing even on shingle or the poorest soil, when chopped and bruised, is a useful winter food for sheep, horses, and cattle.

When the price of corn, inflated by our prolonged wars, had unduly extended the 'margin of cultivation,' a few years of bad harvest stimulated the popular outcry for the abolition of the Corn-laws, and English agriculture, naturally conservative in its methods, has never been able to bear up against the flood of foreign corn that has poured in upon us in a stream of constantly increasing proportions since the commencement of Free Trade. But little has been done, however, to endeavour by the introduction of new crops to meet this difficulty. The most notable changes that have taken place are those which have accompanied the growth of cattle and dairy-farming, and the increased use of oil-cakes and other feeding-stuffs, and of artificial fertilizers. These lie outside our province ; but the introduction of sewage-farming has been accompanied by the cultivation of the ITALIAN RYE-GRASS (*Lolium italicum*, Braun.), which has achieved a great success. Reference has already been made to the utility of MAIZE and SORGHUM as fodder-plants ; and in 1842 Sir Joseph Hooker introduced the TUSSACK-GRASS (*Dactylis cæspitosa*, Forst.,) of the Falkland Islands, of which cattle are very fond, and which has been successfully established in the Hebrides. The most generally useful, however, of new fodder-plants has probably been the PRICKLY CAUCASIAN COMFREY (*Symphytum peregrinum*, Ledeb.) ; and this and other similar plants have had their value greatly enhanced by the adoption during the last few years of the process of ensilage.

---

## PART IX.—MISCELLANEOUS PRODUCTS.

THOUGH our survey has been necessarily a cursory one, the system adopted has left various substances unnoticed, which could not conveniently be classified in any of the divisions adopted, and for the same reasons they can hardly be classed among themselves.

CORK, which is practically entirely the produce of *Quercus Suber*, L., the CORK OAK, a native of Southern Europe, is imported to the extent of several thousand tons annually, in spite of its extreme lightness. It comes mainly from Spain, and is employed not merely for corks for bottles and bungs for casks, but for boot-soles, hat-linings, life-buoys, and various other purposes. Of late years the VIRGIN CORK, or first crop taken off trees ten to fifteen years old, which is furrowed longitudinally, has been imported to ornament ferneries, and for similar purposes. The subsequent removal of the cork every few years, care being taken not to cut the cork-cambium, or growing-layer, does not affect the vitality of the tree, though probably rendering its timber less dense.

AMADOU or, GERMAN TINDER, consists of slices of the fungi *Polyporus fomentarius*, Fries, and *P. ignia-rius*, Fries, beaten out with mallets, and used as a styptic in surgery, for warm underclothing, or, when mixed with saltpetre, as tinder.

DUTCH RUSHES, though imported to a small extent from Holland, are the stems of a British plant *Equisetum hyemale*, L., one of the HORSETAILS, a group allied to the Ferns. Its furrowed stem is so full of silica as to serve to polish wood or metal. It is now mainly superseded by sand, glass, and emery-papers.

The manufacture of bougies and knife-handles from the stalks of the seaweed *Laminaria*, the use of

13

SHOLA-PITH, derived from the Indian *Æschyomene aspera*, L., for hats, and that of the fibro-vascular skeleton of the gourd *Luffa ægyptiaca*, DC., as a bath-towel, are among the more anomalous products. The various INSECT-POWDERS derived from the powdered flowers of *Pyrethrum*, the use of *Quillaia* bark as a hair-wash, and of BAY-RUM for the same purpose, could not strictly be classed under Materia Medica; nor could those extensive industries, the manufacture of acids, such as acetic and oxalic acids from saw-dust; pyrogallic, gallic, and tannic acids from galls; or valerianic acid from potato-spirit.

The long-practised use of the heads of the TEAZLE (*Dipsacus fullonum*, L.) to raise a pile on cloth is in-teresting as a case in which art has failed to supersede a natural product. The cultivated form, with a cylin-drical head, may be derived from the wild species in which the head is ovate. The flower-head is covered with recurved, elastic, hooked scales. Rows of these heads in a frame form part of the gig-mill, or dress-ing-machine, metallic teazles not having proved satis-factory.

The importation of the VEGETABLE IVORY, or COROZO-NUT (*Phytelephas macrocarpa*, Hook.), and the COQUILLA-NUT (*Attalea funifera*, Mart.) for buttons, umbrella - handles, etc., has, since 1843, largely increased, but is likely to be seriously affected, as may various other industries, by what is undoubtedly the most remarkable of these miscellaneous vegetable products of the last half-century, Celluloid. CELLULOID, or PYROXYLINE, is vegetable fibre, cotton rags, paper-waste, or weeds, treated with nitric and sulphuric acids, made plastic with various solvents, especially camphor and castor or linseed oil, and then moulded or dyed in imitation of ivory, tortoise-shell, amber, malachite, coral, or parchment, into artificial teeth or gums, shirt-collars, billiard-balls, or what not ; more durable than ivory, cheaper and more durable than caoutchouc, and, unlike vulcanite, free from sulphur. Though mainly made in the United States, it was invented in England by Mr. A. Parkes in 1885.

Truly, though the introduction of a new resin or fibre, or the invention of a simpler process of manufacture, may not subtend so large a visual angle in the minds of the unthinking as the storming of a citadel or the bombardment of a fleet, when we consider the many strides which our civilization and its attendant comforts have made even during one half-century, owing to the careful observations and skilful application of such men as those to whose names we have had occasion to refer, we must admit that

' Peace hath her victories
No less renowned than war.'

# SYNOPTICAL INDEX.

# GENERAL INDEX.

The names in italics are those of genera mentioned for their uses. The last number after each refers to the Synoptical Index, where the useful species will be found enumerated.

14—2

THE END.

BILLING AND SONS, PRINTERS, GUILDFORD.

Printed in the United States
By Bookmasters